Special Topics in Primatology
Volume 1

Primate Conservation:

The Role of Zoological Parks

No contributors to this volume have received remuneration.
All profits from the sale of *Primate Conservation: The Role of Zoological cal Parks* will go directly to the American Society of Primatologists' Conservation Fund. For more information on the Conservation Fund, please contact a Society officer or visit the web page at:

http://www.asp.org/asp/conservation/

Special Topics in Primatology
Horst Dieter Steklis, Series Editor

Volume 1 **Primate Conservation: The Role of Zoological Parks**
Edited by Janette Wallis

Special Topics in Primatology
Volume 1

Primate Conservation:
The Role of Zoological Parks

Edited by **Janette Wallis**

A Publication of the
American Society of Primatologists

Copyright © 1997   The American Society of Primatologists

Printed in the United States of America
Library of Congress Catalog Card Number: 97-72389
ISBN: 0-9658301-0-1
Cover design, page design, setup, & typography: Janette Wallis

Photo Credits:
Front Cover (top to bottom):
   Mountain Gorilla (*Gorilla gorilla beringei*), © H. Dieter Steklis
   White-handed Gibbon (*Hylobates lar*), © Connie Bransilver
   Cotton-top Tamarin (*Saguinus oedipus*), © Michael Dick
   Golden-crowned Sifaka (*Propithecus tattersalli*), © David Haring
Back Cover (top to bottom):
   Golden Lion Tamarin (*Leontopithecus rosalia*), © Jessie Cohen
   Lion-tailed Macaque (*Macaca silenus*), © Chip Kamber
   Drill (*Mandrillus leucophaeus*), © W.S. Bain
   Bonobo (*Pan paniscus*), © Connie Bransilver
Spine:
   Chimpanzee (*Pan troglodytes*), © Janette Wallis

Library of Congress Catologing-in-Publication Data

Special Topics in Primatology.

Contents: v. 1. Primate Conservation: The Role of Zoological Parks/ editor Janette Wallis.

Includes index
Includes appendix

ISBN: 0-9658301-0-1
1. Primates - Collected Works. I. Wallis, Janette 1954-
II. Title

1997   97-72389

CONTENTS

CONTRIBUTORS

Benjamin Beck
National Zoological Park
Smithsonian Institution
Washington, D.C. 20008-2958
[113]

Gilbert K. Boese
Zoological Society of Milwaukee County
10005 W. Bluemound Road
Milwaukee, Wisconsin 53226
[215]

Mary Bowman
Zoo Atlanta
800 Cherokee Avenue SE
Atlanta, Georgia 30315
[113]

Warren Brockelman
Center for Wildlife Research
Mahidol University,
Rama 6 Road
Bangkok 10400, Thailand
[177]

Katherine Castle
Minnesota Department of Natural Resources
Box 7, DNR Bldg
500 Lafeyette Road
St. Paul, Minnesota 55155
[177]

Cathleen Cox
Los Angeles Zoo
5333 Zoo Drive
Los Angeles, California 90027
[151]

Humberto Giraldo
Roger Williams Park Zoo
1000 Elmwood Avenue
Providence, Rhode Island 02907
[97]

Laurence Gledhill
Woodland Park Zoological Garden
5500 Phinney Avenue North
Seattle, Washington 98103
[131]

Kenneth Gold
Lincoln Park Zoo
2200 N. Cannon Drive
Chicago, Illinois 60605
[43]

Kunkun Jaka Gurmaya
Biologi-FMIPA-UHPAD
JG. Raya Bandung Sumcdang KM 21
Sumedang 45363
Jawa Barat, Indonesia
[177]

Michael Hutchins
American Zoological Association
   Conservation Center
7970-D Old Georgetown Road
Bethesda, Maryland 20814
[29]

John Iaderosa
St. Catherines Wildlife Survival Center
Midway, Georgia 30315
[131]

Charlene Jendry
Columbus Zoological Gardens
9990 Riverside Drive
Powell, Ohio 43065
[199]

Fred W. Koontz
Science Resource Center
Wildlife Conservation Society
185th St. and Southern Blvd.
Bronx, New York 10460
[63]

John Lehnhardt
Disney's Animal Kingdom
P.O. Box 10,000
Lake Buena Vista, Florida 32830-1000
[113]

Donald Lindburg
Zoological Society of San Diego and
Center for Research on Endangered Species
P.O. Box 551
San Diego, California 92112
[131]

Russell Mittermeier
Conservation International
2501 M. Street, N.W., Suite 200
Washington, D.C. 20037
[xi]

Ingrid Porton
St. Louis Zoological Park, Forest Park
St. Louis, Missouri 63110
[83]

Gay Reinartz
Zoological Society of Milwaukee County
10005 W. Bluemound Road
Milwaukee, Wisconsin 53226
[215]

Anne Savage
Roger Williams Park Zoo
1000 Elmwood Avenue
Providence, Rhode Island 02907
[97]

Luis Soto
Proyecto Tití
A.A. 211 Sincelego, Sucre, Colombia
[97]

Tara Stoinski
Zoo Atlanta
800 Cherokee Avenue SE
Atlanta, Georgia 30315
[113]

Jatna Supriatna
Department of Biology
University of Indonesia
Depok 16424, Indonesia
[177]

Ronald Tilson
Minnesota Zoological Gardens
13000 Zoo Blvd.
Apple Valley, Minnesota 55124
[177]

Schwann Tunhikorn
Wildlife Research Division
Royal Forest Department
Raholyothin Road, Bangkhen
Bangkok 10900, Thailand
[177]

Janette Wallis
Oklahoma Biological Survey
University of Oklahoma
111 E. Chesapeake Street
Norman, Oklahoma 73019
[1]

Robert Wiese
Ft. Worth Zoological Park
1989 Colonial Parkway
Ft. Worth, Texas 76110
[29]

Sukie Zeeve
The AZA Madagascar Fauna Group
P.O. Box 2082
Setauket, New York 11733
[83]

Numbers in brackets correspond to the first page of contributor's chapter.

FOREWORD

Over the past several years, conservation of biological diversity has finally taken its rightful place on the global stage. All of our planet's biodiversity is of great importance, but clearly certain groups of species and a small number of particularly rich ecosystems deserve special attention. Among these are the non-human primates: the monkeys, apes, lemurs, lorises, galagos, and tarsiers, that are our closest living relatives and occupy a special position in the imagination of our own primate species, *Homo sapiens*. As research over the past four decades has shown, these animals can teach us a great deal about ourselves and our evolution. Fortunately, the continuing growth of primatology and primate-oriented anthropological research clearly demonstrates that interest in our closest relatives is not likely to wane anytime soon.

Apart from the ways in which primates can enlighten us about ourselves, research on them has intrinsic value and is relevant to research on tropical ecology. About 90% of all primates are found in the tropical rain forests of the world, the richest and most diverse terrestrial ecosystems. Primates occupy a very important role in these habitats. For example, the role of larger rain forest species like the Amazon spider and woolly monkeys as seed dispersers seems to be especially critical in maintenance of forest structure and diversity. Our knowledge of how important primates are in tropical forests grows every year, but is still in its infancy.

Unfortunately, wild populations of nonhuman primates are in trouble in many of the 92 countries where they occur. Some of the most serious problems are in those nations richest in primates, such as Brazil, Peru, Colombia, Madagascar, Indonesia, China, and Vietnam. Primates are threatened by destruction of forests and other natural habitats, by hunting as food (especially severe in West and Central Africa and parts of Amazonia), and by live capture for export (although the impact of this threat has declined considerably in recent years). As a result, of the roughly 250 species of nonhuman primates currently recognized, about half are considered to be of conservation concern by the World Conservation Union (IUCN) Primate Specialist Group. One in five is already in either the endangered or critical category, meaning that they could become extinct in the next couple of decades. Although we will probably make it through this century without having lost a single primate species or subspecies (something that cannot be said for most other major groups of vertebrates), we will enter the next millennium with a large portion of the Order Primates on the edge.

To prevent the extinction of a significant percentage of our primate relatives, we must continue the collection of more and better information. We remain surprisingly ignorant about basic issues of taxonomy and geographic distribution for most primate species, and new taxa continue to be determined - either through taxonomic revisions of museum specimens or discovery in the wild. The most striking examples of new discoveries over the past decade have been in Madagascar and Brazil. In Madagascar, two distinct new species, *Hapalemur aureus* and *Propithecus tattersalli*, were discovered and described in the latter half of the 1980s. A third species, *Microcebus myoxinus*, described over 100 years ago and forgotten, was found to be quite distinct (and also is now the smallest living primate). Other new species will almost certainly be found in Madagascar in the future. The record is even more striking in Brazil, with six new species of primates being discovered and described since 1990 (*Leontopithecus caissara*, *Callithrix nigriceps*, *Cebus kaapori*, *Callithrix mauesi*, *Callithrix marcai*, and *Callithrix saterei*). Discoveries continue in other parts of the world as well, with a new form of colobus uncovered in the Niger Delta of Africa and several new galago taxa determined or discovered in the past few years. In addition to these spectacular discoveries, almost every new field expedition obtains significant information on geographic distribution, range extensions, and behavior.

Despite the growth in conservation activities, there have been major gaps in primatology, one in particular being the lack of quality publications for the general public, e.g., books depicting these amazing and tremendously interesting creatures and capturing their great diversity. When most people think about primates, they think only of chimpanzees, gorillas, baboons, and maybe the occasional capuchin monkey. However, the Order Primates is wonderfully diverse in ways that few people recognize. We of the conservation community are especially interested in portraying this rich diversity and helping to stimulate public concern for endangered primates; top-notch publications play a critical role in the efforts to disseminate our message as widely as possible.

If we are to achieve our goal of maintaining the current diversity of the Order Primates into the next century and beyond, a wide variety of conservation actions will be needed. These include special protection measures for critically endangered species, maintenance of protected areas in regions of high primate diversity and abundance, establishment of new protected areas where needed, reducing or eliminating hunting pressure and live capture for trade, and much more research in all aspects of primate biology in the wild and in captivity. Zoos have played and will continue to play a major role in global strategies for primate conservation. Primates serve as popular exhibit animals in zoos and are excellent ambassadors for the ecosystems that they inhabit, especially the tropical rain forest. As a result, they play a key role in conservation education efforts, an aspect of zoo work that is increasingly recognized as perhaps the greatest contribution that zoos can make to the biodiversity conservation movement as a whole.

Zoos are also focal points for research of many different kinds; technologies to care for, breed, handle and transport primates are more often than not developed first in zoos and then applied in areas of natural habitats. If the natural habitats of primates become increasingly fragmented, application of management techniques in the captive setting will become more and more important. Finally, captive breeding of endangered and critically endangered primates in zoos will continue to have a role. Although it is impossible to keep significant separate colonies of all threatened primates in captivity, breeding of particularly significant flagship species - such as those discussed in this volume - will continue to make an important contribution. Special efforts to conserve critically endangered species, the 35 or so taxa of primates that are likely to be the first to go if appropriate measures are not taken, require very special attention in this regard. In certain cases, reintroduction of captive animals to the wild will be a real option (*e.g.,* golden lion tamarins); in other cases, maintenance of long-term breeding colonies may increasingly be seen as an end in itself.

In any case, there is no doubt that there is an evolving trend toward recognizing that zoos and the wild are not two isolated end points on the conservation spectrum; it is no longer an "us and them" situation. Rather, there is an increasing recognition that a continuum exists between what is done to conserve primates in the zoo setting and what must take place in the wild. More and more people are working in this continuum, helping to bridge gaps that need no longer be a barrier to progress and to work collaboratively to achieve what are certainly mutual goals in primate conservation. This book makes a major contribution to recognizing this continuum and the key role that zoos play in global primate conservation efforts. In so doing, it helps to advance the cause in a very significant way, and I would like to offer my congratulations to the editor and to all the authors in this volume for their outstanding efforts.

Russell A. Mittermeier, Ph.D.
Chairman, IUCN/Primate Specialist Group
and
President, Conservation International

PREFACE

As the inaugural volume in the American Society of Primatologists' book series, *Primate Conservation: The Role of Zoological Parks* is especially relevant. Regardless of field of expertise, all primatologists - at one level or another - are concerned with the conservation of the primates we study.

The idea behind this book can be traced to a long telephone conversation between Joe Erwin and myself, circa 1994. At the time, Joe served as President of the ASP and I chaired the ASP Publications Committee. On that afternoon, we had a rambling discussion about many Society issues, including: the varied aspects of primate conservation; our wish that more zoo personnel were active in the ASP; ideas for fund-raising activities for the ASP; and the pros and cons of the Society becoming a book publisher (an idea Joe had developed some time ago). By the end of the conversation, these four topics all came together. We would make a formal proposal to the Publications Committee to establish a book series to be published by the ASP. The sale of the books would benefit the ASP's future activities, such as those focussing on education, research, and conservation. Concurrently, I organized a symposium for the annual ASP conference that would focus on the zoo community's work in primate conservation and I offered to turn the proceedings of that symposium into the first volume of the book series (with proceeds designated for the Conservation Fund).

In developing this plan, I was fortunate to find several individuals eager to participate in the symposium, and others unable to attend the meeting but interested in contributing to the resulting volume. The symposium was held in June, 1995, at the ASP meeting in Scottsdale, Arizona. Six participants (Cox, Jendry, Lindburg, Reinartz, Savage, and Wiese) spoke to a large and enthusiastic audience; the symposium was successful in showcasing the fine primate conservation work being sponsored by zoos. Some months later, through the diligent work of Erwin, the ASP Board officially approved the development of an ASP book series and the publication of this volume became a more serious endeavor. In addition to chapters contributed by the six symposium participants (and their co-authors), I invited contributions for six additional chapters. The reader will see many familiar names from the zoo and conservation community, mixed with the names of newcomers to the field. Combined, these twelve chapters provide a comprehensive look at the various aspects of zoo-based primate conservation.

I want to give my heartfelt thanks to the contributing authors for making my first book editing experience a pleasant one. I appreciate their dedication to producing a quality volume and their willingness to comply with the reviewers' suggestions and those of my own. I also admire their stamina during my gentle nagging about deadlines.

I thank the members of the 1994-1996 ASP Publications Committee for providing helpful comments during the early planning stages of this book, in particular Jim Moore and Don Lindburg. I am grateful to the many reviewers who provided professional, thorough, and speedy critiques of earlier versions of each chapter. In addition, several individuals donated photographs that, as the reader will see, add immensely to the overall quality of the book.

Because the American Society of Primatologists serves as the publisher of the book series, it is the Editor's responsibility to carry out all phases of production - up to and including production of a camera-ready copy of the book. Thus, for the last year my home office became a publishing house (with my 20 year old cat, Millie, sleeping near the computer in her role as production assistant). Once I completed my part, however, I relied on a printer who would turn the book into a high quality, yet cost-efficient product. Special thanks to Gene Moser and Ed Pham, of Transcript Press, for their professionalism and for making me feel as though I knew what I was doing. Moreover, if this book is in readers' hands by June 26, 1997, I will even forgive Ed and Gene for calling all primates "monkeys."

Finally, I want to thank you, the reader. If you purchased this book, you have made a valuable contribution to the ASP's Conservation Fund. If you incorporate what you learn here into your own work and daily activities, you will help the cause of conservation even further. Keep in mind that this book is not intended to be a critical evaluation of the zoo community. Although several chapters mention the fact that zoos took a relatively long time to develop a conservation ethic (and many zoos are still not involved in primate conservation programs), the purpose of this book is to congratulate zoos for the good work they *are* doing rather than criticize them for what they are *not* doing. With this positive message, the book may better serve to encourage currently active zoos to expand their efforts and to motivate additional zoos to join us in the important mission of primate conservation.

Janette Wallis, Ph.D.
Editor

con•serve (kən-sûrv′) *vt.* **-served, -serv•ing, -serves.** [ME *conserven* < OFr. *conserver* < Lat. *conservare* : *com-* (intensive) + *servare*, to preserve.] **1. a.** To protect from loss, harm, or depletion **b.** To use carefully or sparingly, avoiding waste. – **con•serv′a•ble** *adj.* – **con•serv′er** *n.*

con•ser•va•tion (kən-sûr-vā′shən) *n.* **1.** The act or process of conserving.

Two young gorillas from the Belgian Congo (now Zaire or the Democratic Republic of Congo) are shown minutes after being captured, circa 1931. The pair was brought to the U.S. where they lived for several years in the San Diego Zoo. (Photo by Martin Johnson, courtesy of the *Martin and Osa Johnson Safari Museum, Kansas.*)

FROM ANCIENT EXPEDITIONS TO MODERN EXHIBITIONS: THE EVOLUTION OF PRIMATE CONSERVATION IN THE ZOO COMMUNITY

Janette Wallis

Oklahoma Biological Survey, University of Oklahoma, Norman, Oklahoma

INTRODUCTION

In recent years, zoos in the United States have developed innovative and success-ful primate conservation programs. They started with *ex situ* projects, including breeding programs for endangered species and conservation education for zoo visitors. More recently, they added *in situ* projects, including preservation of natural habitats, sur-veys of primate populations, and helping source country citizens find alternative ways to manage natural resources. Although the zoo community is becoming active in pri-mate conservation, this was not always the case; only during the last 20 years have zoos made conservation a top priority.

This chapter traces the history of conservation awareness in zoos and describes the factors leading to the need for primate conservation action. Many of these issues have been well-documented over time. Publications from the late 19th Century and early 20th Century provide rich, often poignant, prose on the subject of nonhuman primates. Therefore, this chapter uses direct quotations from the literature to illustrate the evolution of nonhuman primate status in the zoo world: from ancient expeditions to modern exhibitions.

THE ACQUISITION OF ZOO ANIMALS

The first zoos in the United States opened their gates in the late 1860s to mid-1870s. Based on success in European zoos, nonhuman primates were considered much desired animals for exhibit in the U.S. Thus, zoo managers endeavored to collect pri-mate species from around the world – sparing little expense and wasting little time.

Early expeditions to collect wild primates used a wide variety of techniques, some aimed at tricking the animal into a trap, others aimed at capturing infants and juveniles by first killing their mothers and other adults. No thought was given at that time to the long-term consequence of removing individuals from their group nor removing large numbers of animals from an already limited population.

> A favorite native method of hunting is with birdlime, which is a mucilage made from the gum of a tree. Birds and monkeys are captured in birdlime smeared in the limb of trees; they stay in it until some one goes up and pulls them out. Another way of capturing small monkeys is by means of a sweetened rag in a bottle. The bottle is covered with green rattan and tied to a tree. The monkey puts his hand through the neck and grabs the rag. He cannot pull his hand out while it is doubled up with the rag in it, and he hasn't sense enough to let go. There he sticks, fighting with the bottle, until the hunter comes along and, by pressing the nerves in his elbow, forces him to open his hand and leave the rag for the next monkey. [Mayer, 1920, p. 45]

Bates [1910] wrote of collecting the white uakari (*Brachyurus calvus;* now known as *Cacajao calvus,* and regarded as a subspecies of the bald uakari): "individuals are obtained alive by shooting them with the blow-pipe and arrows tipped with diluted *urari* poison.... A pinch of salt, the antidote to the poison, is then put in its mouth, and the creature revives. The species is rare, even in the limited district which it inhabits. When an independent hunter obtains one, a very high price is asked, these monkeys being in great demand for presents to persons of influence down the river" [p. 329]. The bald uakari is now listed as vulnerable to extinction [World Conservation Union (IUCN), 1996].

> Monkeys and the smaller apes are caught in specially-constructed traps. These are, as a rule set up near the waterhole which the creatures are in the habit of frequenting. The traps are of various kinds, and designed according to the particular species of monkey it is desired to capture. In the case of the bigger apes only the young are taken. This generally means that the mother, and sometimes both the parents, have to be shot before the baby is removed. [Shepstone, 1932, pp. 53]

Often many animals were killed in the effort to obtain only one or two specimens. In the early days, there was a fine line between the skills and motivation of a "hunter" and those of a "collector." One hundred years ago, Garner [1896] published one of the first detailed descriptions of gorilla and chimpanzee behavior and anatomy, used as an educational text by many in the early zoo community. In that impressive volume, the frontispiece displayed the author posing with his shotgun. One cannot imagine such an author photograph appearing in a modern book on great apes. Before widespread use of the shotgun, however, other methods were used to kill adult apes:

> As they do in the case of most dangerous animals, the native collectors hunt orang-outangs by killing the mother and taking the young. The weapon they most often use, except when they have guns, is the

blow-pipe, which, in the hands of an expert, is to be despised. It is a long, slender tube, measuring from six to eight feet, made from a single joint of a rare bamboo. The tube is allowed to dry and harden and is wrapped tightly with rattan. The darts, which are about the size of a steel knitting-needle, are made from the mid-ribs of palm leaves, and at one end there is a small conical butt, which fits tightly in the bore of the pipe. A small nick is made in the shaft of the dart just below the point, and the end is coated with a deadly poison made from the sap of the upas tree and another species of the genus *Ipo*. When the dart strikes, the end breaks off and remains in the wound; the poison acts rapidly, first paralyzing, then killing the victim. [Mayer, 1920, p. 131]

When a female orangutan was killed, her infant was captured live and kept in the village until the opportunity arose to sell it. Wrote one animal collector, "More than once in a Borneo village have I seen a woman nursing a baby on one breast and an infant orang on the other. Always there is that hope that a trader will come along to buy the ape" [Buck, 1930, p. 97].

On occasion, a U.S. zoo requested collection of an adult ape, which required a different set of tactics than those used to capture young apes. For example, to catch two large orangutans in Sumatra, local workers cut away all trees surrounding a sleeping platform, effectively treeing the apes. The platform tree was felled and the apes were netted as they hit the ground [Mayer, 1920]. During the struggle, one orangutan reportedly killed a Sumatran helper, grasping him by the throat and whipping him through the air to break his neck. Though they were successfully collected, both orangutans died before being exhibited in a zoo. In a later expedition, Mayer [1924] captured a young orangutan in a baited trap affixed to a tree. During the process of moving the cage to the ground, however, two adult orangutans – reportedly trying to defend and free the captured animal – attacked two of Mayer's men. In the ensuing struggle, both humans died and both adult orangutans were shot dead. Still, the expedition was successful for Mayer; in the following ten days the team trapped three gibbons, four baby orangutans, one proboscis monkey, ten "black monkeys," eighteen long-tailed monkeys, twenty-two pig-tailed monkeys, and various other mammals and reptiles [Mayer, 1924].

Well-known animal collector, Frank Buck [1930], described his life as a collector of wild animals for zoos, circuses and dealers. Over an 18 year period, he brought a number of Asian primates into the United States, including 52 orangutans, 31 gibbons, and over 5,000 monkeys of different species. Later, Buck bragged, "I am mighty proud of the fact that in all my twenty-five years of experience with animals I have devoted myself entirely to live animals rather than big game hunting, museum collecting, or anything that necessitated killing them" [1936, p. 260]. On the contrary, though Buck was well-known for the phrase "Bring 'em Back Alive," this was not an entirely accurate description of his work. He may have left the source country with live animals, but an enormous number died en route to the United States. For example, he described the capture of an adult male orangutan, weighing 250 lbs. The locals gave the orangutan native gin (*arric*), causing him to pass out drunk and making him easy to capture. Although the actual collection occurred without incident, the animal died aboard ship while destined for a U.S. zoo. A similar fate was suffered by proboscis monkeys Buck purchased from a Malay trader. The male died during transit and, soon after, the female gave birth to an infant that lived only a few days. After the infant's death, the mother stopped eating, succumbing soon thereafter. All three deaths occurred before the animals reached their zoo destination [Buck, 1930].

Buck and many other animal dealers often avoided personal involvement in the capture of wild primates, preferring instead to purchase primates from the local market. The following account comes from Gerald Durrell, who spent many years collecting animals for British zoos. On this occasion, he traveled to Africa to collect animals for his own personal zoo:

> In the Cameroons, monkeys are one of the staple items of diet, and, as there are no enforced laws covering the number that are shot or the season at which they are shot, it is natural that a vast quantity of females carrying young are slaughtered. The mother falls from the trees with the baby still clinging tightly to her body, and in most cases the infant is unhurt. Generally the baby is then killed and eaten with the mother; occasionally the hunter will take it back to his village, keep it until it is adult, and then eat it. But when there is an animal collector in the vicinity, of course, all these orphans end up with him, for he is generally willing to pay much more than the market price for the living animal. So at the end of two or three months in a place like the Cameroons you generally find that you are playing foster parent to a host of monkeys of all shapes and ages. [Durrell, 1960, pp. 130-131]

Similar stories were told by Marlin Perkins [1982], who purchased a mandrill, a baboon, and a green monkey in one Cameroon marketplace. Though Durrell, Perkins and others may have had good intentions by "rescuing" the orphaned monkeys, the willingness to pay a high price for live animals clearly supported the market, increasing the horrible practice of killing adult primates to collect their young for sale.

Very rarely in the literature, one may trace the entire captive life of wild-caught primates. Such is the case of *Mbongo* (initially named *Congo*) and *Ngagi,* two gorillas brought to the U.S. in 1931. Martin and Osa Johnson tell the story of the apes' capture in the "Alumbongo Mountains" (Mt. Tshiaberimu district) of the Belgian Congo (now Zaire) [Johnson, 1931]. Johnson's team used a method similar to that used by Mayer (above) – chasing the pair into a tree and then cutting down all surrounding trees. When their tree was felled, the startled young gorillas were netted and tied to a pole. The photo on the face page of this chapter shows the pair just minutes after their capture. Each weighed more than 100 lbs. at the time of collection, indicating that they were approximately 6 years of age. At the time, they were the largest gorillas ever collected alive; Johnson noted the pair was caught "without causing injury to man or beast, which was something to boast about" [1931, p. 209].

The young gorillas survived their capture admirably and were transported to the United States. After a short stay in New York, they were eventually sold to the San Diego Zoo for $11,000 [Stott, 1993]. Johnson accompanied his prized apes to their final destination, but expressed grave concern over their fate:

> Now that I have my animals safely in America I am sorry that I brought them home. To see *Ingagi* and *Congo* imprisoned behind iron bars and steel netting in a space far too small, when only a few months ago they had the wide space of the Congo in which to roam, make me regret their capture. [Johnson, 1931, p. 278]

> Fate already has spun the wheel for the members of our once happy menagerie. My hand can no longer stay its course. Again I say, I regret having brought these animals to civilization. I will never send another into captivity. [p. 281]

Though initially considered mountain gorillas *(Gorilla gorilla beringei)*, due to the high elevation of their home range, the pair was later re-classified as eastern lowland gorillas *(Gorilla g. graueri)* [Groves & Stott, 1979; and see Groves, 1967].

After *Mbongo* and *Ngagi* became residents in the San Diego Zoo, their story was continued through the writings of Belle Benchley [1949]. Their arrival in December of 1931 was a major boost to the zoo. They lived well into adulthood, eventually weighing 580-620 lbs. As a means of enriching the gorillas' environment, zoo staff provided them with straw each night to use as bedding material. Unfortunately, this kind gesture on the part of zoo management had fatal consequences for the gorillas. *Mbongo* (Congo) died in March, 1942, *Ngagi* in January, 1944 [Benchley, 1949; J. Ogden, personal communication]. Autopsy reports revealed coccidiodal granuloma, a growth of the fungus, *Coccidioides immitis*, which was found in the straw bedding material. The condition, today called "Valley Fever," is confined to populations living in the arid southwestern United States [Benchley, 1949; Stott, 1993; and see Morbidity and Mortality Weekly Report, 1994]. Thus, although they beat the odds, surviving longer in a zoo than other apes at that time, the pair succumbed in captivity to a disease that does not even exist in their natural habitat.

ZOO MANAGEMENT AND ANIMAL HEALTH CARE

As described above, many primates captured for zoos did not live to see their intended destination. Johnson [1931] estimated that, in his time, not more than one in 25 animals survived the journey. Even if they survived the journey, most lived only a short time in captivity due to a lack of expertise in animal husbandry. In fact, due to poor knowledge of health care and diet (said to include beer for apes), many primates imported by British zoos died while in quarantine or very soon after being displayed to the public. The first ape viewed by London Zoo visitors, a Gambian chimpanzee collected in 1835, survived for only six months in the zoo [Blunt, 1976].

> The only drawback with these creatures (chimpanzees) is that they are liable to succumb to lung troubles. It is the same with the gorilla. The more important zoological gardens have spent fortunes on young gorillas, but they seldom live longer than a few months. [Shepstone, 1932, pp. 53-54]

During two trips to the Belgian Congo, Burbridge [1928] captured 8 mountain gorillas and killed one. The Burbridge team's standard method of capture included wrestling a young ape to the ground, putting a bag over its head, then beating it into submission. Only two of the eight captives survived to reach their intended destination. One of these, *Miss Congo*, was studied by Robert Yerkes at his laboratory in Florida. Of those that died, one was stung to death by army ants and five died of various diseases contracted from their human captors. Burbridge cited the important lesson learned from this experience:

> Until now the fallacy of placing captive gorillas in intimate association with humans, from whom they contract colds and influenza, has not been recognized. It has always been supposed that the society of humans instead of being a menace was a necessity, and that when the young gorilla changed masters it immediately pined away and died of grief, when in reality the sickness was due entirely to another cause. Fresh air, exercise, and proper diet are the essentials necessary for their welfare. The inhuman practice of placing animals in solitary confinement so prevalent in menageries is especially hard on the gorilla, who should have a young chimpanzee for a cage mate, one whose active play would give him ample exercise. [Burbridge, 1928, p. 254]

Little progress had been made in the area of animal husbandry by the time U.S. zoos established great ape exhibits. In the Cincinnati Zoo Guide, published in 1923, General Manager Sol Stephan confirmed that it was exceedingly difficult to keep chimpanzees alive in captivity for more than three or four years, yet he claimed that the Cincinnati Garden "endeavors to have one or more chimpanzees in its collection at all times" [p. 8]. Indeed, the Cincinnati Zoo has had a long history of housing chimpanzees or their congeners, bonobos (*Pan paniscus*) (M. Dulaney, personal communication).

Whereas early experience indicated that most primate species were ill-suited for captive living (or, more accurately, that zoos were ill-suited for caring for primates), the practice continued and, gradually, zoo officials improved their skills in caring for these charismatic creatures. Still, zoos drew criticism for holding wild animals in captivity. Johnson [1931] remarked, "I have no quarrel with zoos. I like to visit them and study the animals as well as anyone. But I must say that it is cruel to keep the animals penned up the way they are in most such places in the civilized world. The poor dumb brutes cannot complain of their quarters which are usually far too small" [p. 279]. William Hornaday, early Director of the New York Zoological Society, responded to such criticisms by citing the importance of the quality of care:

> Many persons sincerely believe that all wild animals in captivity 'pine' for their native jungles; but this is not by any means the case with animals which are caught quite young, and are kept in comfortable quarters. Of course, any poor creature caged by a mean man in an evil way, in a small cage, is miserable; and it does pine hard for more room, good air, good light and decent society....There are just the same differences in keeping of captive animals that there are in the keeping of wives and children; and I ask all real humanitarians not to lose sight of this evident fact. [Hornaday, 1925, p. 206]

His "politically incorrect" statement regarding women and children notwithstanding, Hornaday correctly pointed to the need for care and consideration in the keeping of wild animals in captivity. In time, this attitude would prove a key component in assuring long lives for zoo inhabitants, and, eventually, help to phase out the importation of primates from the wild. Until then, however, countless primate deaths occurred unnecessarily during collection, transport, quarantine, and even while on exhibit in zoos.

MEANWHILE, BACK IN THE FOREST...

While zoo primates were suffering through humans' lessons in animal husbandry, their wild counterparts were also suffering at the hands of humans. Although the collection from the wild for zoos had a minimal impact on animal endangerment, throughout history, there have been a number of environmental pressures (including those imposed by humans), on the natural habitats of nonhuman primates. Although these pressures were not always well documented in the literature, by all accounts their impact has increased steadily and substantially in the last century. It is generally agreed there are three primary areas of risk to wild primates: habitat destruction, illegal live trade, and hunting [see Gibbons, 1995].

Deforestation and Primate Habitat Destruction

The overwhelming threat to wild primate populations is the worldwide destruction of rain forests. More than 90% of all nonhuman primates live in the tropical forests of Asia, Africa, and South and Central America [Mittermeier et al., 1986]. The World Resources Institute [1990] reported that 20.4 million hectares of tropical forests are falling annually to deforestation. Fully 20% (450 million hectares) of all tropical natural forest cover was lost from 1960 to 1990; Asia lost almost one-third of its tropical forest covering during that period, whereas Africa and Latin America each lost approximately 18% [Food and Agriculture Organization (FAO), 1995]. In Africa, almost 70% of forest changes in the 1980s occurred via degradation of closed forest to open and fragmented forests, suggesting that rural population pressure (through subsistence farming, grazing, and wood extraction for fuel wood and building materials) is the primary agent behind forest change [FAO, 1995; World Resources Institute et al., 1996].

Unfortunately, it is difficult to trace documented changes in forests over the last century, although occasionally in the early literature there is a hint of concern or a description that can be compared to conditions today. For example, famed explorer Stanley [1890] described in intricate detail a lush central African forest, estimated to cover 321,057 square miles, in the (then) Congo Free State. The area is now known as Zaire's Aruwimi Basin. Within this, the Ituri Forest has been designated as one of six primary tropical African forests requiring urgent action for primate protection [Oates et al., 1982].

As early as 1932, Shepstone noted the main causes which have led to the destruction of wild animals in Africa, citing:

> ...the extent and intensity of cultivation and the consequent reduc-
> tion in areas suitable to the habits of wild animals. As civilization
> advances the wild-life must necessarily disappear. The area over
> which the creatures once roamed is curtailed as land is brought un-
> der cultivation. The animals have consequently to retire further afield
> or be destroyed. [Shepstone, 1932, p. 226]

Only in recent years have the topics of "conservation," "reforestation," and "habitat destruction" appeared with any frequency in the literature. A search of the WorldCat online library database, which covers over 36 million books and documents dating from the 17th century to the present, identified 1063 books specifically addressing the subject of "deforestation." Of these, only 20 (1.9%) were published before 1940 and most of those focussed only on forests in the United States. Indeed, one notable domestic forest conservation message was given in a U.S. Presidential address. President Theodore Roosevelt, in 1907 [as cited in Lundeberg and Seymour, 1910, p. 455], asked, "Shall we continue the waste of our natural resources, or shall we conserve them? There is no other question of equal gravity now before the nation." Such a strong conservation philosophy is to be admired in a world leader. Unfortunately, while he championed the urgent protection of America's natural resources, Roosevelt occasionally participated in the destruction of Africa's natural resources; the President was an avid hunter of big game and African monkeys (see below).

Live Capture for Commercial Trade

Due to the eventual success of captive management and breeding and the heightened awareness of nonhuman primates' precarious conservation status, the capture of primates for U.S. zoo collections has virtually ceased in recent years. As stated earlier,

Figure 1. Young male chimpanzee (*Pan tro-glodytes schweinfurthii*) confiscated in Kigoma, Tanzania. Along with five other young chimpanzees and more than 40 African Grey parrots (*Psittacus erithacus*), he was apparently smuggled out of Zaire, taken by boat across Lake Tanganyika, and was reportedly destined for a zoo in the Middle East. His swollen abdomen is indicative of worm infestation and malnutrition. (Photo by Janette Wallis.)

past collection for U.S. zoos had only a minor effect on wild population numbers. However, large scale live capture – for mostly illegal commercial trade – still continues to be a problem for many primate species. Animals captured for commerce usually end up in non-accredited zoos or as privately owned pets (Figure 1). Indeed, the keeping of nonhuman primates as pets is not a new phenomenon; it is a long-standing practice, even in habitat countries:

> Although monkeys are now rare in a wild state near Pará, a great number may be seen semi-domesticated in the city. The Brazilians are fond of pet animals. Monkeys, however, have not been known to breed in captivity in this country. I counted, in a short time, thirteen different species, whilst walking about the Pará streets, either at the doors or windows of houses, or in the native canoes. Two of them I did not meet with afterwards in any other part of the country. [Bates, 1910, p. 50]

Very often, the illegally captured nonhuman primate does not live long enough to make money for its captors. Domanalain, a self-described "poacher turned gamekeeper," published detailed 'confessions' of his years as an animal trafficker. He claims to have exported over 300 gibbons to various countries around the world and provided a lengthy account of the horrible conditions the animals suffer along the way. Often, if an intended recipient refused the cargo, the animal was abandoned at the airport by the animal trafficker, to be disposed of by airport officials.

> I feel that here we have reached the height of absurdity: we capture animals in the wild, we submit them to incredible ordeals, many die of hunger, of sickness, of injuries... and, when the survivors finally reach their destination after all this suffering, we gas them without pity. [Domanalain, 1977, translated by M. Barnett, p. 40]

Nichol [1987], also a reformed collector of animals for zoos, cites the importance of conservation awareness in the very recent past: "Twenty-five years ago there was no talk of conservation, natural habitats weren't being exploited and not a murmur was heard about the depletion of natural resources" [p. 1]. Travelling in India in those days, Nichol asked a local man the whereabouts of monkeys that were previously so abundant. "Oh," the main said with a shrug, "they're in your country" [p. 1]. Nichol believes that the Convention on International Trade of Endangered Species (CITES), aimed at curtailing illegal trade, may have actually made animal trafficking more attractive. When the animals became harder to obtain, their value increased, and a new generation of animal smugglers was born. Regardless of Nichol's view, CITES has indeed helped to regulate primate trade. Any species listed on Appendix II needs an export permit from the source country and any species listed on Appendix I needs both the export permit and an import permit from the country of destination [Mack & Mittermeier, 1984].

Hunting: Nonhuman Primates as Food, Medicine, Ornamentation and Curios

Food. Although we often cite the senseless killing of adult nonhuman primates to collect the young for illegal trade, in most cases, it is the killing – and eating – of the adult primates that is the primary purpose of the hunt (Figure 2); the orphaned infant is simply a resulting "bonus." The young may be sold as a pet or kept long enough to grow into a substantial meal in its own right. In fact, it may be remarkable that any nonhuman primate species survives in great number today; throughout the literature, almost without exception, the taste and palatability of primate meat is described positively:

> The gastronomical possibilities of a monkey probably occur to only a very few of the millions who gaze at him through iron bars. Yet the natives of South America consider the flesh of certain species of monkeys to be superior to that of almost any other animal. The meat of the larger forms, the howling, spider, and woolly monkeys, is especially sapid. [Cutright, 1940, p. 154]

> The Indians are greatly elated when one or more of the large monkeys is killed. They eat everything but the skin and break open the bones for the marrow they contain. Some tribes go so far as to wash out the contents of the stomach, which, with the addition of water, is said to make a palatable drink. This consists for the most part of

Figure 2. Bushmeat for sale in a marketplace in Moanda, Gabon, (western equatorial Africa). Along with a leopard, there are a grey-cheeked mangabey (*Lophocebus albigena*) (hanging in the upper left corner) and a black colobus monkey (*Colobus satanas*). The appendages in the foreground are those of a mandrill (*Mandrillus sphinx*). (Photo by Janette Wallis.)

only leaves and fruits that have been acted upon by the enzyme pepsin, so that the idea of swallowing such a beverage is not so nauseating as it seems on first thought. [p. 155]

In describing an expedition of the Amazon, Up de Graff [1923] recalled: "I selected a young *coto* (howling monkey) from the bag, skinned it and roasted it on the embers. Prepared in this way, they are just as good eating as any other game I have tasted except their cousin the *maquisapa* (spider monkey), whose equal, in my estimation, cannot be found on any table" [p. 229]. Indeed, early explorer Bates [1910] wrote the following of what he called a *coaitiá*, or spider monkey (*Ateles marginatus*):

I thought the meat the best flavoured I had ever tasted. It resembled beef, but had a richer and sweeter taste. During the time of our stay in this part of the Cuparí, we could get scarcely anything but fish to eat, and as this diet ill agreed with me, three successive days of it reducing me to a state of great weakness, I was obliged to make the most of our Coaitiá meat.... Nothing but the hardest necessity could have driven me so near to cannibalism as this, but we had the greatest difficulty in obtaining here a sufficient supply of animal food. [p. 215]

Today, *A. marginatus,* endemic to Brazil, is officially listed as an endangered species [World Conservation Union, 1996a].

Alfred Russell Wallace [1889], who shares with Darwin the distinction of first proposing the theory of evolution by natural selection, wrote the following account after his companion shot a non-identified South American monkey: "The poor little animal was not quite dead, and its cries, its innocent-looking countenance, and delicate little hands were quite childlike. Having often heard how good monkey was, I took it home, and had it cut up and fried for breakfast. There was about as much of it as a fowl, and the meat something resembled rabbit, without any very peculiar or unpleasant flavor" [p. 29]. Hornaday [1925] later made a similar remark, "Most monkeys are, like squirrels, too smelly to eat; but the teetee of South America, roasted, is far better than any squirrel-meat that I ever tried" [p. 167].

Clearly, the frequency with which South American monkeys appear on the menu has taken its toll. The following account provides startling estimates of how many woolly monkeys were routinely killed in the Upper Amazon:

> In the forest the Barrigudo (*Lagothrix lagotricha*) is not a very active animal; it lives exclusively on fruits, and is much persecuted by the Indians, on account of the excellence of its flesh as food. From information given me by a collector of birds and mammals, whom I employed, and who resided a long time amongst the Tacuna Indians, near Tabatinga, I calculated that one horde of this tribe, 200 in number, destroyed 1,200 of these monkeys annually for food. The species is very numerous in the forests of the higher lands, but, owing to long persecution, it is now seldom seen in the neighborhood of the larger villages. It is not found at all on the Lower Amazon. [Bates, 1910, p. 335]

It is unclear to which subspecies of woolly monkey (*L. lagotricha)* Bates refers, but both the *L.l cana* and *L. l. poeppigii* are now listed as vulnerable to extinction in Brazil [World Conservation Union, 1996].

Most literary references to eating primates in Asia are for medicinal purposes (see below). However, Schaller [1993] noted that the slow loris (*Nycticebus coucang*) is a local favorite best cooked with lemon leaves.

There are many accounts in the literature of great apes being hunted for food across Africa. The following passages refer to mid-19th Century West African views of chimpanzees, in the region now known as Liberia, and gorillas, in the region now known as Gabon. Note the very similar statements regarding our relationship to apes:

> It is a tradition with the natives generally here, that they (chimpanzees) were once members of their own tribe; that for their depraved habits they were expelled from all human society, and, that through an obstinate indulgence of their vile propensities they have degenerated into their present state and organization. They are, however, eaten by them, and when cooked with the oil and pulp of the palm

nut, considered a highly palatable morsel. [Savage & Wyman, 1843-44, p. 385]

The natives of the interior are very fond of the meat of the gorilla and other apes. Gorilla-meat is dark red and tough. The sea-shore tribes do not eat it, and are insulted by the offer of it, because they suspect some affinity between the animal and themselves. In the interior some families refuse to eat gorilla-meat from the superstitious belief explained elsewhere, that at some time one of their female ancestors has brought forth a gorilla. [Du Chaillu, 1861, pp. 400-401]

From a journal entry dated August 24th, 1870, Livingstone described the practice of eating chimpanzees in eastern Zaire (then called Congo Free State): "The flesh of the feet is yellow, and the eagerness with which the Manyuema devour it leaves the impression that eating *sokos* was the first stage by which they arrived at being cannibals: they say the flesh is delicious" [Waller, 1875, p. 223].

The eating of nonhuman primates continues today, although it is illegal in most areas. A recent report in the London Times described an illegal shipment of (unspecified) monkey meat from Equatorial Guinea into Spain [Varadarajan, 1996]. The problem was revealed when baggage handlers at the airport detected the stench of rotting meat. The shipment was destined for sale to Madrid's Equatorial Guinean immigrant community, for whom the meat is considered a prized delicacy.

In Gabon, the eating of primates was implicated in a recent outbreak of the Ebola virus, which killed twenty people in the town of Mayibout. The exact origins of the virus are not known, but it is presumed to have a natural host in the forest, through which it infects primates. Many of those who died in the Gabon Ebola outbreak feasted on chimpanzee meat in recent days. Pygmy hunters in the area reported increasing encounters with dead gorillas and chimpanzees in the forest, apparently killed by a mysterious affliction [French, 1996].

The rapid commercialization of the wildlife bush meat trade in Africa has gained a great deal of attention during the last two years [see Pearce, 1996]. For the African apes, the bush meat trade may be an even greater threat than habitat destruction. In 1995, an estimated 800 gorillas were slaughtered for meat in one territory alone. Some conservationists estimate as many as a ten gorillas and chimpanzees may be taken from the forests of West and Central Africa for the cooking pot every day [Pearce, 1996]. A campaign led by Kenyan conservationist Karl Ammann and the World Society for the Protection of Animals (WSPA) has alerted political leaders in the European Union and African-Pacific-Caribbean nations to the crisis. A joint EU-APC Congress meeting in Namibia voted unanimously to sanction the European timber industry and African governments that developed the infrastructure to support the bush meat trade. They have demanded that officials police the roads and logging towns to stop the slaughter of great apes and other threatened wildlife. In response, the government of Cameroon held a conference in April, 1996, confirming the adverse impact of forest exploitation on wildlife. American Anthony Rose, working on behalf of the people of Cameroon and the entire region, is organizing North American resources to help reverse the devastation. As part of this organized effort, Rose is urging a number of North American zoos to form partnerships with wildlife conservation and education NGOs in the countries where the great ape bush meat threat is most severe. Meetings have been held with staff at the Los Angeles Zoo, Columbus Zoo, and Lincoln Park Zoo to develop cooperative ventures to address the bush meat problem in Cameroon and in other African countries (A. Rose, personal communication).

Medicine. In many source countries, nonhuman primates are used in preparation of traditional medicines. Schaller [1993] described the use of primates in Asian traditional medicines and suggested that the rare Yunnan golden monkey may soon become extinct if its brains continue to be used for medicinal purposes. A number of stories abound in Africa about the healing powers of ape brains. In Sierra Leone, for example, chimpanzee brains smeared over a man's body is said to cure infertility (R. Nisbett, personal communication).

The Nilgiri langur and the lion-tailed macaque, both threatened with extinction, are killed by local people in the Western Ghats for their flesh. Soup made from these monkeys is highly regarded as a rejuvenent and also thought to cure a great number of common ills like cold, influenza, flatulence, lassitude, and low blood pressure [Sheshadri, 1969]. The broth of gibbon bones is said to cure epilepsy [Schaller, 1993].

In Asia, the great disparity in wealth creates a dual problem regarding hunting of nonhuman primates. With increasing affluence in parts of China, people can afford to purchase valuable nonhuman primates for use in medicines. There is now a growing market for monkeys smuggled from Vietnam into China for this purpose (A. Eudey, personal communication). Conversely, in areas of great poverty in Asia, nonhuman primates are more in danger of being hunted for food. In either case, many Asian primate species are at risk.

Fetishes. In 19th Century Liberia, fetishes were made from herbs and grasses beaten in a mortar and mixed with clay, pot black, and oil. "This is spread over a goat's horn and left to dry. A chain, a deer's horn filled with medicine, and two bells are attached to this horn. The idol is wrapped in monkey skin, leaving the chain and bells hanging out. The whole is tied on a man's back and worn in a war with the devil-doctor's assurance that, while he continues to wear it, he is perfectly safe, even in the front of the battle, and that no musket ball can pierce his skin" [McAllister, 1896, p. 249].

As recently as 1991, a Sierra Leone infant was observed wearing an amulet made of leather and a chimpanzee incisor. This item was tied around the waist of the child to protect it and give it power over others in his cohort (R. Nisbett, personal communication).

> In the Qing Dynasty (1641-1911), only high officials were allowed to wear clothing made of golden-monkey skins. In the 1930s, a golden monkey pelt would bring the large sum of $200 in the local fur market. Every Qiang used to have a personal guardian spirit that manifested itself in objects or other creatures, especially in the golden

monkey. Priests often carried the skull, liver, fingernails, and other monkey parts wrapped in white sheets of paper in remembrance of the tribe's sacred books. [Schaller, 1993, pp. 27-28]

Ornamentation and Curios. Poaching is by no means confined to killing for meat or medicine: "From most of the tens of thousands of animals killed by poachers in Eastern Africa today, the trophies are cut out, most of them to be used for trivial or ignoble purposes, and the meat simply left to rot" [Huxley, 1960, p. 18]. Although this practice typically affects large game, such as the rhinoceros and elephant, the hides of colobus monkeys, for example, have found a ready market across Africa and Europe. Lundeberg and Seymour [1910] discussed the African travels of former U.S. President Theodore Roosevelt and cited the collection of the black and white colobus monkeys:

> But its fur, unfortunately, is a much desired article of trade, and therefore the animal is pursued and its numbers greatly diminished by European and native hunters, who armed with breech-loading rifles have almost exterminated it in many of its favorite haunts, so that a few years ago it took an experienced African hunter 5 days to secure three specimens for a European museum. Our former President may, therefore, consider himself fortunate in having bagged one of these rare animals. In a not far distant future it, no doubt, will be too late, for the war of extermination has been carried on even to the remotest mountain forests to satisfy the demand for the fur of the colobus. An African traveler found hundreds of skins ready for shipment to Europe by Greek and Indian traders, where they are used as trimmings and linings of ladies' winter coats. A missionary told him that he himself had hunted eighty animals within a month to sell their fur, for which he received from one to two dollars apiece. While its fur was 'in fashion' hundreds of thousands of the animals were exported to Europe to satisfy a passing fancy. Before the European invasion the natives hunted the colobus only because its fur was used by their warriors to adorn their ankles. [pp. 228-229]

The hoolock or white-browed gibbon (*Hylobates hoolock*) of India is hunted by the tribesmen in Assam; the fur is used as part of the adornment of men and women [Sheshadri, 1969]. Kingsley [1897] wrote of a local woman in west Africa wearing a necklace made of sixteen gorilla canine teeth, slung on a pineapple fiber string.

Perhaps even more disturbing than using primates for food, medicine, or fetishes is the practice of collecting trophies or curios. Fossey [1983] discussed the gruesome collection of the hands of mountain gorillas from Rwanda. This irreverence for non-human primates is not new. Up de Graff [1923] described the brisk demand for curios in Panama: shrunken heads of either human or monkey made to order and sold for $25 each (1923 US$).

Endangered Species: The Case of Gorillas

As a result of the pressures described above, many primate species have suffered devastating reductions in their numbers. Ongoing re-assessment of species classification makes it difficult to provide an exact number, but there are roughly 220 to 236 species of nonhuman primates alive today. The 1996 IUCN Red List of threatened animals shows 13 primate species as critically endangered, 28 endangered, 52 vulnerable, and 41 at some degree of lower risk to extinction [World Conservation Union, (IUCN), 1996]. Thus, perhaps as many as 60% of all primate species are in need of special protection.

As the largest primate species, gorillas have held a prominent place in natural history literature since they were first described in the mid-1800s [Savage & Wyman, 1843-44; see review in Schaller, 1963]. Early reports provided brief details of their behavior, but they were often viewed as more valuable dead than alive. To make a thorough examination of their anatomy and physiology, a great number of specimens were collected for museums and laboratories. As described earlier, live capture for zoos was a frequent occurrence in the early part of this century, though the captives often died before reaching their destination. Thus, the human fascination with gorillas led to a rapid and drastic decline in their numbers.

Some of the earliest accounts of western lowland gorillas (*Gorilla g. gorilla*) referred to their being present by the thousands in the area now known as Gabon [Du Chaillu, 1861]. No official census data are available for Gabon, but this subspecies is found over a relatively wide area of Africa (including Angola, Cameroon, Central African Republic, Congo, Equatorial Guinea, Gabon, and Nigeria). Although their numbers are estimated to be as high as 111,000 [Harcourt, 1996], in light of heavy habitat destruction and the bush meat trade (see above), a more realistic figure may be as low as 50,000 (A. Meder, personal communication). The subspecies is listed as endangered, with status in Cameroon and Nigeria designated as "critically endangered" [World Conservation Union, 1996].

The gorillas of east Africa are in even more peril. Early in this century, many expeditions set out to track and collect gorillas in the Belgian Congo. Among the early collectors was Carl Akeley, whose mountain gorilla specimens are on display in the Museum of Natural History in New York. His interest and respect for gorillas led him eventually to turn away from museum collecting, working instead to protect gorilla habitat. Though he was admired for this work, not all of his colleagues agreed with his assessment of the gorilla's status.

> Believing that gorillas were due to become extinct if shooting of them continued, Akeley worked hard to induce the Belgian government to reserve these mountains as a gorilla sanctuary. I now know, after finding the large numbers which we did on our last safari and in nine different districts, that there is no possible chance of them becoming extinct, no matter how many may be shot. [Johnson, 1931, p. 160]

> In my estimate for the Belgian government, I reported that there are
> about 2,000 gorillas in the Kivu district, and I recommended that
> the present gorilla preserve be extended to include other mountain
> fastness, enlarging the gorilla sanctuary from two hundred and fifty
> to approximately five hundred square miles. There is now no threat-
> ened extinction of the gorilla in this section by white or native hunt-
> ers. But there is a spotted menace, a potent factor too, in the leopard,
> who destroys numbers of young animals. [Burbridge, 1928, p. 246]

A similar figure of 2,000 gorillas was also calculated by Johnson [1931] for the same area, specifically including Mounts Mikeno, Visoki (Visoke), Sabino (Sabinyo), and Karisimbi, all now known as part of the Virunga Volcanoes. Johnson's remark that gorillas will endure "no matter how many are shot" is particularly absurd in view of current census figures. Today, there are no more than 600 mountain gorillas (*G. g. beringei*) remaining on Earth (A. Meder, personal communication).

Further west, near the mining town of Alimbongo, Johnson made the following calculations for what he referred to as the "Alumbongo Mountains":

> After doing a little figuring, my deductions were that this gorilla
> country extended 40 miles from east to west and 47 miles from north
> to south. This would make an area of more than 1800 square miles
> of ape country. Judging from what I had seen of the gorilla packs
> and the numbers in them, I estimated that there were at least 20,000
> gorillas in the Alumbongo range. [Johnson, 1931, p. 196]

Obviously, Johnson's calculations were not scientifically obtained, but most of today's gorilla experts agree that his estimate was within reason for that time. Today, there are a mere 50 gorillas found in the Alumbongo region (E. Sarmiento, personal communi-cation). Estimates for the entire eastern lowland gorilla subspecies (*G. g. graueri*) range from 4,000 (A. Meder, personal communication) to 10,000 [Harcourt, 1996].

> I am now convinced that the gorilla population of the Belgian Congo
> is many, many times larger than ever before supposed. I am equally
> certain that these animals are in no danger of extermination. I know
> that the Mikeno gorillas are not hunted by natives, now that the Parc
> National Albert has been established. Whether they were hunted be-
> fore, I have no means of knowing. We certainly saw no evidence of
> gorillas being slain by the inhabitants of the Alumbongo Mountains.
> The skulls these people brought to me, I am sure, were those of
> animals that had died from natural causes. [p. 212]

We cannot fault Johnson for trying to be optimistic, but his implying that one can rule out the possibility of a gunshot wound to the chest, simply by examining a skull, is a naïve suggestion at best. Today, we know there have been – and continue to be – many pressures on gorillas in East Africa, such as ever-increasing habitat destruction, killing for trophies [Fossey, 1983; see Jendry, this volume], and hunting for food (described earlier).

Whereas humans must take the blame for the decline in gorilla populations, ironically, the gorillas' own behavior may have played a small role in their demise. When approached by strange humans, a wild chimpanzee will tend to flee immediately – high in the trees, if possible. This makes chimpanzees somewhat difficult to capture or kill. A male gorilla, on the other hand, will stand up to a threat, beating his chest and eventually charging his enemy [Du Chaillu, 1861]. This apparent act of "bravery" leads to his certain death, however, for his great strength is no match for a gun.

> Fortunately, the gorilla dies as easily as man; a shot in the breast, if fairly delivered, is sure to bring him down. He falls forward on his face, his long, muscular arms outstretched, and uttering, with his last breath, a hideous death-cry, half roar, half shriek, which, while it announces his safety to the hunter, yet tingles his ears with a dreadful note of human agony. It is this lurking reminiscence of humanity, indeed, which makes one of the chief ingredients of the hunter's excitement in his attack of the gorilla. [Du Chaillu, 1861, p. 358]

There are many such accounts in the early literature regarding the stalking and killing of gorillas. Though the writers may have provided gruesome detail, it was often tempered with a note of regret. In a search of the early accounts, Benchley [1949] made a similar observation:

> This is a strangely consistent thing. In every account, no matter how ruthlessly and with what determination the hunter sought his prey and slew him, there lurks in the telling of the tale a strange feeling of regret, almost of shame, at having accomplished his purpose; and always there is the implication of the strong fascination the beast has held for the man. And usually, in so many plain words, the hunter says that it is only his eagerness to assist in building up knowledge of the beast or the necessity to save a human life that has prompted the killing. Each claims that at the last minute, when he has come face to face with the gorilla, he has felt the greatest reluctance to destroy a creature which, after days or hours of hunting, in which it has been near him, seeing him without being visible, hearing him while moving soundlessly itself, watching and knowing that it was being sought, has finally been forced to meet its pursuer when cornered and has stood erect so like a man that it has become almost impossible to destroy it in cold blood. Its death has left the hunter feeling like a murderer. [p. 184]

The following passage is a perfect example of Benchley's point and may be seen as a harbinger of conservation awareness. Written by a collector of museum specimens, it is a compelling description of how gorillas have suffered at the hands of humans. The author recorded these words after seeing the slain body of a western lowland gorilla in southeastern Nigeria:

> I shall never quite forget the emotions that this sight conjured up inside me. I had always been taught to think of the gorilla as the very essence of savagery and terror, and now there lay this hoary old vegetarian, his immense arms folded over his great pot belly, all the fire gone from his wrinkled black face, his soft brown eyes wide open beneath their long straight lashes and filled with an infinite sorrow. Into his whole demeanor I could not help but read the tragedy of his race, driven from the plains up into the mountains countless centuries ago by more active ape-like creatures - perhaps even our own forebears; chevied hither and thither by the ever-encroaching hordes of hairless shouting little men, his young ones snatched by leopards, his feeding grounds restricted by farms and paths and native huntsmen. All around him was a changing world against which he bellowed his defiance to the end, rushing forward to meet the bits of lead and gravel blasted at him by his puny rival. [Sanderson, 1937, pp. 181-182]

RESPONSE FROM THE ZOO COMMUNITY

From the time they first began, zoos have undergone a constant learning process. Initially, they learned how to keep animals alive in captivity, addressing proper nutrition, health, handling, and other facets of animal husbandry. Zoo managers have recognized that the failure to reproduce an environment that is at least functionally equivalent to that of the wild will inevitably result in the loss of many forms and patterns of natural behavior. Therefore, zoos learned about the animals' habits and behavior, so they could make informed decisions about social housing, space allocation, and functionality of the exhibits, while addressing the species' psychological and physiological requirements. They worked to improve exhibit design, for the benefit of the animals as well as the zoo visitor. Thus, all changes were aimed at improving conditions for the animals and improving the zoo's image to the public. Though zoos have always had their critics, much of this criticism has been taken to heart and incorporated into an ongoing self-examination [see various chapters in Norton et al., 1995]. Through the years, zoos have asked some very serious questions:

- Do zoos exist merely to entertain the public or can they serve to educate the public, as well?
- How can zoos respond to the pressures reducing animal populations in the wild?

As described in the following sections (and in this entire volume), these two questions are being answered through creative and innovative programs, making the zoo community an important factor in the field of primate conservation.

The Zoo's Focus: Entertainment vs. Education

In the early days of zoos, a major focus was on public entertainment. Much discussion concerned the "performance" potential of various species and, in some cases, zoo personnel concerned themselves more with training animals than learning about "normal" animal behavior [see Hornaday, 1922; Stephan, 1923]. While serving as the Director of the New York Zoological Park, Hornaday [1922] published a book that focussed on this practice. He provided a photograph captioned as "Christmas at the Primates' House," showing chimpanzees (designated in the legend as "with large ears") and orangutans (designated as "small ears"), fully dressed, sitting in chairs at a table [p. 42]. A decorated Christmas tree was nearby, and zoo patrons looked on as the animals ate with forks and spoons [see Gold, this volume, for photo and further discussion of this topic].

Although a small amount of "performance" work is still conducted in zoos, using such mammals as dolphins and elephants, reputable zoos no longer dress and train anthropoid apes for public entertainment. Instead, the trend is to exhibit primates in naturalistic settings, while educating the public about the animals' natural behavior and biology. The last several years has seen major changes in primate zoo exhibit design and the accompanying educational graphics [reviewed by Lindburg & Coe, 1995; Gold, this volume].

Critics of zoos may suggest that some modern zoo projects, such as the Orangutan Language Project at the National Zoo's Think Tank Exhibit, are continuing the practice of making apes "perform" for the public; having an orangutan use a computer may be just as "anthropocentric" as having a chimpanzee ride a bicycle. However, there are major differences between these modern projects and the ape shows of the past. The animals in the Think Tank live in a social setting, are trained using positive reinforcement, and perform on a voluntary basis. Most important, these exhibits serve to educate the public more than to entertain them; they highlight the cognitive capacity of the apes and present a much more positive image of primates. Although the Think Tank is not focussed on conservation education, per se, one may argue that when zoo visitors learn more about the intelligence, sentience, and creativity of great apes, they may gain a better appreciation for the intrinsic value of our primate relatives and develop a stronger conservation ethic [see Gilbert, 1996, for details of the Think Tank exhibit].

The Growth of Conservation Awareness in U.S. Zoos

As in all areas of enlightenment, it took time for conservation awareness to grow in the zoo community. By the 1950s and 1960s it was clear that zoos were changing their focus. Well-known Swiss Zoologist, Heini Hediger [1959, 1960], wrote extensively about wild animals in captivity, placing major emphasis on the need for research and conservation programs in zoos. Gerald Durrell [1960] pointed to the need for captive breeding programs: "In my opinion zoological gardens all over the world should have as one of their main objectives the establishment of breeding colonies of such rare and threatened species. Then, should what seems to be inevitable happen and the animal become extinct in the wild state, at least it is not lost completely" [p. 4]. This sentiment was echoed by Attenborough who, after a zoo expedition to Madagascar, made the following remarks concerning the endangered lemurs on the island:

> It is urgently important that steps should be taken to save [endangered primates] and that the existing restrictions should not be relaxed. Zoos must play their part in this by taking some of these animals and providing them with spacious enclosures, the prime function of which is not to exhibit them unrelentingly to the public but to give them the right conditions under which they can breed.
>
> But while this is a wise policy to pursue as a precaution against total extermination, the safest protection the animals can receive is the rigorous conservation of their native habitat. [Attenborough, 1961, p. 155]

By the mid-1970s, wildlife conservation was a major concern in the U.S. zoo community, both in thought and in deed. The IUCN Zoo Liaison Group for the American Association of Zoological Parks and Aquariums (AAZPA) recommended a Zoo Conservation Committee be established to promote more effective participation by zoos in conservation work [Schomberg, 1972]. In 1981, the AAZPA (now known as the AZA) established the Species Survival Plans©, a coordinated propagation program to sustain some of the world's most endangered species [Conway, 1995; Hutchins & Conway, 1995; see details in Weise & Hutchins, this volume]. Through careful planning, the SSP assures genetic diversity within captive populations, and for a very few primate species, the breeding program can potentially help replenish populations in the wild [see Cox, this volume; Stoinski et al., this volume]. Most importantly, the SSP program allows zoos to breed nonhuman primates for exhibit; for most species, no further collection from wild populations is needed [see Lindburg et al., this volume].

In addition to SSPs, the AZA has developed a number of advisory groups and committees to coordinate zoo programs nationwide and, in some cases, worldwide [see Weise and Hutchins, this volume]. Development of various regional Fauna Interest Groups (FIGs) have made a huge impact on primate conservation [see Zeeve and Porton, this volume, for discussion of the Madagascar Fauna Group].

Individual zoos have worked to improve primate exhibits and provide the visitor with conservation education in the form of graphics, lectures, and newsletters. To help this process, zoos are hiring well-educated staff. In fact, they are encouraged to employ a Curator of Science, "sufficiently cogent in the methods of science to be able to read the literature and to extract from it that which is locally applicable" [Lindburg,

1993, p. 318]. This, and other such plans, will help improve all areas of zoo management, aid in information exchange between institutions, and foster new zoo-based research and conservation efforts.

Modern zoo conservation work is not confined within the walls of a zoo. As detailed in this volume, many *in situ* projects have been developed or financially supported by the U.S. zoo community. In this way, zoo personnel can support conservation projects in habitat countries, work to halt population decline, and perhaps repay a debt owed to nonhuman primates displayed in our zoos.

Although this book highlights the efforts of U.S. zoos, the drive to promote conservation work in the zoo community is an international effort [see discussion by Gibbons, 1995]. The *World Zoo Conservation Strategy* was issued in 1993 [World Zoo Organization and World Conservation Union, 1993]. It points to the potential of modern zoos having the largest conservation network on Earth. The Strategy encourages zoos around the world to expand their involvement in conservation:

- By actively supporting the conservation of populations of endangered species and their natural ecosystems, while continuing work with captive breeding programs;
- By offering support and facilities to increase scientific knowledge that will benefit conservation, citing the importance of having staff psychologists, zoologists, and veterinarians; and
- By promoting an increase in public awareness of the need for conservation.

In addition, the IUCN's Species Survival Commission's Conservation Breeding Specialist Groups (CBSG) has developed significant processes to assist in establishing conservation priorities. For example, a Population Habitat Viability Analysis (PHVA) focuses on a single species or subspecies or population thereof. On the basis of detailed analyses, the PHVA process presents recommended strategies to address the species' recovery and survival [see Tilson et al., this volume, for discussion of gibbon PHVAs].

Perhaps the best way to illustrate the evolution of conservation awareness in zoos is to examine the words of leaders in the zoo community. Compare the following two statements by the Director of the New York Zoological Park, in 1922, to the Director of Zoo Atlanta, in 1995:

> There are several reasons why chimpanzees predominate on the stage, and why so few performing orang-utans have been seen. They are as follows: 1) The orang is sanguine, and slower in execution than the nervous chimpanzee, 2) The feet of the orang are not good for shoes and biped work, 3) The orang is rather awkward with its hands, and finally, 4) There are fully twice as many chimps in the market. [Hornaday, 1922, p. 82]

(early in the next century...) Collectively, zoos and aquariums will be the most active, well-funded, and effective conservation organizations in the world. Zoos and aquariums will be routinely allied with research and teaching institutions throughout the world, providing a multitude of opportunities to advance basic and applied zoo biology. Zoos and aquariums will be the acknowledged leaders in the successful marketing of a universal conservation ethic and a cooperative network for environmental problem-solving. As mainstream conservationists, zoos and aquariums will inspire individuals, groups, and corporations to work actively on behalf of wildlife and wildlife habitat locally and globally. [Maple, 1995, p.28]

This direct comparison may not be entirely fair. Certainly, Hornaday's words did not represent the attitude of all his contemporaries and, sadly, Maple's vision of a zoo's potential is not shared by all directors in the modern zoo community. However, the important zoos of today are actively incorporating research and conservation aspects into their stated purpose [Hutchins & Conway, 1995; Hutchins et al., 1996]. It is not enough simply to create new exhibits for primate species. It is not enough simply to participate as an SSP member institution. Zoos must go a step further to develop programs aimed at ensuring a future in the wild for the species they keep in captivity [see Mallinson, 1995].

CONCLUSIONS

As this chapter describes, many pressures on primates have led to an urgent need for primate conservation action. Some of the preceding discussion is disturbing to read; detailed accounts of pain and death make us uncomfortable. But such are the realities facing us today. We often treat this issue too delicately. We use the terms "disappearing" and "vanishing" in reference to endangered species as if, for example, habitat destruction causes a species to simply evaporate. The process is not so quick and simple, however. The clearing of trees causes slow death due to starvation, lack of reproduction due to insufficient nutrition and stress, crowding and fighting due to too many animals competing for limited resources. Primates are not "disappearing." They are being evicted, captured, and killed. Because humans are responsible for most pressures on primate populations, humans can and should work to lesson these pressures.

Joining with university researchers and conservation organizations, the zoo community has taken up the challenge to address the primate conservation crisis. Even the harshest critic of zoos must acknowledge the progress made in recent years. For a review of zoo-based primate conservation success stories, one need look no further than in the present volume. The commitment and hard work of zoo professionals across the U.S. has made an impressive difference for a number of primate species and additional projects are currently being developed. If we are to succeed in curbing habitat destruction and loss of nonhuman primate populations, it will require a cooperative and creative spirit and will greatly benefit from the support and leadership of the zoo community.

ACKNOWLEDGMENTS

I wish to thank Conrad Froehlich, Director of the Martin and Osa Johnson Safari Museum, for generously granting permission to reprint the photograph appearing on the face page of this chapter and sharing material from the museum archives. In addition, Gary Varner, Angela Meder, Esteban Sarmiento, Dominique Vallet, Tammie Bettinger, Anthony Rose, Jackie Ogden, John Hart, Travis Pickering, and the Frank Buck Zoological Society facilitated access to information otherwise difficult to find.

REFERENCES

Attenborough, D. ZOO QUEST TO MADAGASCAR. London, Lutterworth Press, 1961.

Bates, H.W. THE NATURALIST ON THE RIVER AMAZONS. (Popular Edition), London, John Murray, 1910.

Benchley, B.J. MY FRIENDS, THE APES. Boston, Little, Brown and Company, 1949.

Blunt, W. THE ARK IN THE PARK: THE ZOO IN THE NINETEENTH CENTURY. London, Hamish Hamilton, 1976.

Buck, F. BRING 'EM BACK ALIVE. New York, Simon and Schuster, Inc., 1930.

Buck, F. ON JUNGLE TRAILS. Philadelphia and New York, Frederick A. Stokes Company, 1936.

Burbridge, B. GORILLA: TRACKING AND CAPTURING THE APE-MAN OF AFRICA. London, George G. Harrap & Company, 1928.

Conway, W. Zoo conservation and ethical paradoxes. Pp. 1-9 in ETHICS ON THE ARK: ZOOS, ANIMAL WELFARE, AND WILDLIFE CONSERVATION. B.C. Norton; M. Hutchins; E.F. Stephens; T.L.Maple, eds. Washington, D.C., Smithsonian Institution Press, 1995.

Cutright, P.R. THE GREAT NATURALISTS EXPLORE SOUTH AMERICA. New York, MacMillan Co. 1940.

Domalain, J.-Y. THE ANIMAL CONNECTION: THE CONFESSIONS OF AN EX-WILD ANIMAL TRAFFICKER. [Translated from French by M. Barnett]. New York, William Morrow & Co., 1977.

Du Chaillu, P. EXPLORATIONS AND ADVENTURES IN EQUATORIAL AFRICA. London, John Murray, 1861.

Durrell, G. A ZOO IN MY LUGGAGE, New York, The Viking Press, 1960.

Food and Agriculture Organization of the United Nations (FAO). FOREST RESOURCES ASSESSMENT 1990: GLOBAL SYNTHESIS. FAO Forestry Paper 124, Rome, FAO, 1995.

Fossey, D. GORILLAS IN THE MIST. Boston, Houghton Mifflin, 1983.

French, H.W. An African forest harbors vast wealth and peril. NEW YORK TIMES, April 3, 1996.

Garner, R. L. GORILLAS AND CHIMPANZEES, London, Osgood, McIlvaine & Co., 1896.

Gibbons, E.F., Jr. Conservation of primates in captivity. Pp. 485-502 in CONSERVATION OF ENDANGERED SPECIES IN CAPTIVITY: AN INTERDISCIPLI-

NARY APPROACH. E.F. Gibbons, Jr.; B.S. Durrant; J. Demarest, eds. New York, State University of New York Press, 1995.

Gilbert, B. New ideas in the air at the National Zoo. SMITHSONIAN 27(3):32-43, 1996.

Groves, C.P. Ecology and taxonomy of the gorilla. NATURE 213(5079):890-893, 1967.

Groves, C.P.; Stott, K.W., Jr. Systematic relationships of gorilla from Kahusi, Tshiaberimu and Kayonza. FOLIA PRIMATOLOGICA 32:161-179, 1979.

Harcourt, A.H., Is the gorilla a threatened species? How should we judge? BIOLOGICAL CONSERVATION 75:165-176, 1996.

Hediger, H. WILD ANIMALS IN CAPTIVITY. New York, Dover Publications, 1950.

Hediger, H. MAN AND ANIMAL IN THE ZOO: ZOO BIOLOGY. New York, Delacorte Press, 1969.

Hornaday, W.T. THE MINDS AND MANNERS OF WILD ANIMALS. New York, Charles Scribner's Sons, 1922.

Hornaday, W.T. A WILD ANIMAL ROUND-UP. New York, Charles Scribner's Sons, 1925.

Hutchins, M.; Conway, W.G. Beyond Noah's Ark: the evolving role of the modern zoological parks and aquariums in field conservation. INTERNATIONAL ZOO YEARBOOK 34:117-130, 1995.

Hutchins, M.; Wiese, R.; Willis, K. Why we need captive breeding. Pp. 77-86 in PROCEEDINGS OF THE AMERICAN ZOO AND AQUARIUM ASSOCIATION REGIONAL CONFERENCE. Wheeling, West Virginia, American Zoo and Aquarium Association, 1996.

Huxley, J. THE CONSERVATION OF WILD LIFE AND NATURAL HABITATS IN CENTRAL AND EAST AFRICA. Paris, UNESCO. 1960.

Johnson, M. CONGORILLA. New York, Brewer, Warren, and Putnam, 1931.

Kingsley, M.H. TRAVELS IN WEST AFRICA. London, MacMillan and Company, 1897.

Lindburg, D.G. Editorial: Curators and applied science. ZOO BIOLOGY 12:317-319, 1993.

Lindburg, D.G.; Coe, J. Ark design update: Primate needs and requirements. Pp. 553-569 in CONSERVATION OF ENDANGERED SPECIES IN CAPTIVITY: AN INTERDISCIPLINARY APPROACH. E.F. Gibbons, Jr.; B.S. Durrant; J. Demarest, eds. New York, State University of New York Press, 1995.

Lundeberg, A.; Seymour, F. THE ROOSEVELT AFRICAN HUNT AND WILD ANIMALS OF AFRICA. New York, D. B. McCurdy, 1910.

Mack, D.; Mittermeier, R.A., eds. THE PRIMATE TRADE. Washington, World Wildlife Fund-U.S. and TRAFFIC (USA), 1984.

Mallinson, J.J.C. Zoo breeding programmes: Balancing conservation and animal welfare. DODO, JERSEY WILDLIFE PRESERVATION TRUST 31:66-73, 1995.

Maple, T.L. Toward a responsible zoo agenda. Pp. 20-30 in ETHICS ON THE ARK: ZOOS, ANIMAL WELFARE, AND WILDLIFE CONSERVATION. B.C. Norton;

M. Hutchins; E.F. Stephens; T.L.Maple, eds. Washington, D.C., Smithsonian Institution Press, 1995.

Mayer, C. TRAPPING WILD ANIMALS IN MALAY JUNGLES. London, Asia Publishing Company, 1920.

Mayer, C. WILD BEASTS I HAVE CAPTURED. Garden City, New York, Doubleday, Page & Company, 1924.

McAllister, A. A LONE WOMAN IN AFRICA: Six years on the Kroo Coast. New York, Eaton and Mains, 1896.

Mittermeier, R.; Oates, J.F.; Eudey, A.A.; Thornback, J. Primate conservation. Pp. 3-72 in COMPARATIVE PRIMATE BIOLOGY, VOL. 2A: BEHAVIOR, CONSERVATION, AND ECOLOGY. G. Mitchell; J. Erwin, eds. New York, Alan Liss, 1986.

Morbidity and Mortality Weekly Report (MMWR). Update: coccidioidomycosis — California, 1991-1993. MORBIDITY AND MORTALITY WEEKLY REPORT (NE8), 43(23):421-423, 1994.

Nichol, J. THE ANIMAL SMUGGLERS. New York, Facts on File Publications, 1987.

Norton, B.G.; Hutchins, M.; Stevens, E.F.; Maple, T.L., eds. ETHICS ON THE ARK: ZOOS, ANIMAL WELFARE, AND WILDLIFE CONSERVATION. Washington, D.C., Smithsonian Institution Press, 1995.

Oates, J.F.; Gartlan, J.S.; Struhsaker, T.T. A framework for planning rain-forest primate conservation. IPS NEWS (1):4-9, 1982.

Pearce, J. Slaughter of the apes. SWARA: EAST AFRICAN WILD LIFE SOCIETY 19(1):18-23, 1996.

Perkins, M. MY WILD KINGDOM. New York, E.P. Dutton, Inc., 1982.

Sanderson, I. ANIMAL TREASURE, New York, The Viking Press, 1937.

Savage, T. S.; Wyman, J. Observations on the external habits and characters of the *Troglodytes niger* and on its organization. BOSTON JOURNAL OF NATURAL HISTORY 4:362-376 and 377-386, 1843-1844.

Schaller, G.B. THE MOUNTAIN GORILLA: ECOLOGY AND BEHAVIOR. Chicago, University of Chicago Press, 1963.

Schaller, G.B. THE LAST PANDA. Chicago, University of Chicago Press, 1993.

Schomberg, G. IUCN Zoo Liaison. Pp. 68-71 in PROCEEDINGS OF THE AMERICAN ASSOCIATION OF ZOOLOGICAL PARKS AND AQUARIUMS ANNUAL CONFERENCE. Wheeling, West Virginia, American Association of Zoological Parks and Aquariums, 1972.

Seshadri, B. THE TWILIGHT OF INDIA'S WILD LIFE. London, John Baker Publishers, 1969.

Shepstone, H.J. WILD BEASTS TO-DAY. New York, The MacMillan Company, 1932.

Stanley, H.M. IN DARKEST AFRICA. New York, Charles Scribner's Sons, 1890.

Stephan, S.A. CINCINNATI ZOO GUIDE. Cincinnati, Ohio, Cincinnati Zoological Park Association, 1923.

Stott, K. Letter to Conrad Froehlich, Director, Martin and Osa Johnson Safari Museum, 24 Februay, 1993.

Up de Graff, F. W. HEAD HUNTERS OF THE AMAZON: Seven years of exploration and adventure. Garden City, New York, Garden City Publishing Co., Inc., 1923.

Varadarajan, T. Monkey meat raises a stink at airport. THE LONDON TIMES, 23 May, 1996.

Wallace, A.R. TRAVELS ON THE AMAZON AND RIO NEGRO. (2nd Edition) London, Ward, Lock & Co., Ltd., 1889.

Waller, H. THE LAST JOURNALS OF DAVID LIVINGSTONE. New York, Harper & Brothers, 1875.

World Conservation Union (IUCN). IUCN RED LIST OF THREATENED ANIMALS. Covelo, California, Island Press, 1996.

World Resources Institute (WRI). WORLD RESOURCES 1990-1991: A GUIDE TO THE GLOBAL ENVIRONMENT. New York, Oxford University Press, 1990.

World Resources Institute; United Nations Environmental Programme; United Nations Development Programme; World Bank. WORLD RESOURCES 1996-1997: A GUIDE TO THE GLOBAL ENVIRONMENT. New York, Oxford University Press, 1996.

World Zoo Organization (IUDZG); World Conservation Union (IUCN/SSC/CBSG). THE WORLD ZOO CONSERVATION STRATEGY: THE ROLE OF ZOOS AND AQUARIA OF THE WORLD IN GLOBAL CONSERVATION. Chicago, Chicago Zoological Society, 1993.

Dr. Janette Wallis is a Research Scientist with the Oklahoma Biological Survey. She has more than twenty years' experience studying chimpanzees in both captive and field settings and formerly served as the Coordinator of Chimpanzee Research at Gombe Stream Research Centre (Tanzania). Dr. Wallis has been active with the American Society of Primatologists' (ASP) Conservation Committee since 1992 chaired the ASP'S Task Force on Private Ownership of Primates, and has recently joined the AZA's East Africa Faunal Interest Group.

An adolescent male chimpanzee (*Pan troglodytes*), sits in his lushly landscaped exhibit at the Tulsa Zoo, where he lives with his father, mother, and seven other chimpanzees. He is chewing on a food-coated bone, part of a research project that focussed on bone modification resulting from chimpanzee mastication. (Photo by Janette Wallis.)

THE ROLE OF NORTH AMERICAN ZOOS IN PRIMATE CONSERVATION

Robert J. Wiese and Michael Hutchins

Executive Office & Conservation Center, American Zoo and Aquarium Association, Bethesda, Maryland

INTRODUCTION

The American Zoo and Aquarium Association (AZA) and its 180 member institutions are taking dramatic conservation action on behalf of many primate species. Through a coordinated system designed to facilitate cooperation, zoo and aquarium staff are joining forces with one another and with other similar organizations and individuals in this global effort. Member institutions and their staff are encouraged to maximize their conservation effectiveness through this voluntary program. The AZA has Advisory Groups for prosimians, Old World monkeys, New World monkeys, gibbons, and great apes. These expert committees identify and prioritize primate taxa for which zoo populations are appropriate, recruit studbook keepers to manage population records, help identify research and conservation objectives for priority taxa, and select taxa for Species Survival Plan© (SSP) status. The AZA is moving toward managing the populations of all captive primates held in participating North American zoos. The AZA SSP has recently evolved into a truly holistic conservation program. SSPs implement conservation action through: (1) public education; (2) basic and applied research (including the development of relevant technologies); (3) captive breeding for reintroduction, when appropriate and necessary; and (4) direct support for field conservation.

THE AZA CONSERVATION AND SCIENCE NETWORK

The American Zoo and Aquarium Association's (AZA's) Conservation and Science (C&S) network provides an effective framework for its 180 member institutions to cooperate on a number of conservation and research projects. From 1991 to 1995, zoo-based scientists and their university collaborators produced more than 285 publications that focused specifically on primates. They also initiated, developed, or participated in more than 100 primate conservation and research projects each year during that same period [Hutchins et al., 1991; Wiese et al.,

1992, 1993; Bowdoin et al., 1994; Swaringen et al., 1995]. There are five basic components of the AZA C&S Program that facilitate inter-institutional collaboration within North America. They are Studbooks, Species Survival Plans© (SSPs), Taxon Advisory Groups (TAGs), Fauna Interest Groups (FIGs) and Scientific Advisory Groups (SAGs) [Wiese & Hutchins, 1994].

Studbooks

A studbook is a computerized record of the captive population history of a species. Whereas SSPs, TAGs, FIGs and SAGs are specialized committees composed of several experts from both within and outside the zoo profession, studbook databases are typically maintained by a single, dedicated individual. Each animal is given a unique identification number linked to several types of critical information, including capture date, birth date, death date, parents, siblings, offspring, and moves to various institutions. Studbooks can be regional (i.e., covering participating North American institutions) or international (covering participating institutions worldwide). Regional studbook keepers apply to and are formally recognized by the AZA Wildlife Conservation and Management Committee. International studbook keepers are recognized by the IUDZG/The World Zoo Organization, following regional approval. Many primates held by non-AZA accredited institutions and even some held in private collections are also tracked by these studbooks. AZA currently maintains studbooks on 53 species of primates (Table 1); these studbooks include virtually all primates housed in AZA member institutions. Studbook data are the basis of organized, scientifically-managed captive breeding programs and are critical for proper genetic management. They are also valuable for estimating life history parameters, such as average litter size, infant mortality, and interbirth interval – all necessary for proper demographic management.

Taxon Advisory Groups

Taxon Advisory Groups (TAGs) are expert committees that provide leadership and communication for cooperative zoo-based programs on related taxa. While TAGs vary greatly in how their membership is selected, members are drawn from AZA institutions, universities, industry, state and federal government, and can include private citizens with relevant expertise. AZA studbook keepers and Species Survival Plan© coordinators are automatic members of the TAG. Related species typically share similar husbandry and conservation issues or occupy similar exhibit space in zoos. TAGs focus on issues shared by several related taxa [Hutchins & Wiese, 1991; Hutchins et al., 1995]. All nonhuman primate species are now covered under the umbrella of five AZA TAGs, including prosimians, Old World monkeys, New World monkeys, gibbons, and great apes.

Because captive space and other resources are limited, not all species can receive the focused attention needed for their conservation. Therefore, hard choices must be made to prioritize which species are allocated the limited space, human,

TABLE 1
PRIMATE STUDBOOKS
MANAGED BY AZA MEMBER INSTITUTIONS
I=International Studbook; R=Regional (North American) Studbook

PROSIMIANS
Aye-aye (*Daubentonia madagascariensis*) (I)
Galago (Galaginae) (R)
Lemur, Black (*Eulemur macaco macaco*) (I)
Lemur, Blue-eyed (*Eulemur macaco flavifrons*) (I)
Lemur, Brown (*Eulemur fulvus*) (R)
Lemur, Coquerel's Mouse (*Mirza coquereli*) (R)
Lemur, Fat-tailed Dwarf (*Cheirogaleus medius*) (R)
Lemur, Lesser Mouse (*Microcebus murinus*) (R)
Lemur, Mongoose (*Eulemur mongoz*) (R)
Lemur, Ring-tailed (*Lemur catta*) (R)
Lemur, Black-and-White Ruffed (*Varecia v.*
     variegata) (I)
Lemur, Red (*Varecia variegata rubra*)(I)
Loris, Slow (*Nycticebus coucang*)
Loris, Pygmy Slow (*Nycticebus pygmaeus*)
Loris, Slender (*Loris tardigradus*)
Potto (*Perodicticus potto*) (I)
Tarsier, Phillipine (*Tarsius syrichta*)
Tarsier, Western (*Tarsius bancanus*)
Sifaka (*Propithecus* spp.) (R)

NEW WORLD MONKEYS
Marmoset, Pygmy (*Callithrix pygmaea*) (R)
Marmoset, Geoffrey's (*Callithrix geoffroyi*) (R)
Monkey, Goeldi's (*Callimico goeldii*) (I)
Monkey, Black howling (*Alouatta caraya*) (I)
Monkey, Owl (*Aotus* spp.) (R)
Monkey, Spider (*Ateles* spp.) (R)
Monkey, Titi (*Callicebus* spp.) (R)
Monkey, Woolly (*Lagothrix lagotricha*) (R)
Saki, White-faced (*Pithecia pithecia*) (R)

Tamarin, Bi-colored (*Saguinus b. bicolor*) (I)
Tamarin, Cotton-top (*Saguinus oedipus*) (I)
Tamarin, Emperor (*Saguinus imperator*) (I)
Tamarin, Geoffroy's (*Saguinus geoffroyi*) (R)
Tamarin, Golden Lion (*Leontopithecus rosalia*
     rosalia) (I)

OLD WORLD MONKEYS
Langur, Francois' (*Presbytis francoisi*) (R)
Langur, Spectacled (*Presbytis obscura*) (R)
Macaque, Japanese (*Macaca fuscata*) (R)
Macaque, Lion-tailed (*Macaca silenus*) (I)
Colobus, Black and White(*Colobus guereza*) (R)
Mangabeys (*Cercocebus* spp.) (R)
Monkey, Diana (*Cercopithecus diana*) (R)
Monkey, De Brazza's (*Cercopithecus neglectus*) (R)
Monkey, Patas (*Erythrocebus patas*) (R)
Baboon, Hamadryas (*Papio hamadryas*) (R)
Drill (*Mandrillus leucophaeus*) (I)
Mandrill (*Mandrillus sphinx*) (R)

GIBBONS
Gibbon, Siamang (*Hylobates syndactylus*) (R)
Gibbon, White-cheeked (*Hylobates concolor)* (R)
Gibbon, White-handed (*Hylobates lar*) (R)

GREAT APES
Bonobo (*Pan paniscus*) (R)
Chimpanzee (*Pan troglodytes*) (R)
Gorilla, Lowland (*Gorilla gorilla gorilla*) (R)
Orangutan (*Pongo pygmaeus*) (R)

and financial resources. The primary role of the TAG is to prioritize species for cooperative conservation efforts by AZA institutions and develop a Regional Collection Plan (RCP) [Hutchins et al., 1995]. Using a wide range of criteria, such as current size of the captive population, ability to care for the species, degree of endangerment, public appeal, and potential to act as a flagship species, the TAG selects species for cooperative management programs. To help identify such priorities AZA Primate TAGs work closely with the Primate Specialist Group of IUCN/ The World Conservation Union's Species Survival Commission. These recommendations are then passed along to participating institutions for use in institutional collection and conservation planning. By 1997, all primate TAGs will have completed their initial Regional Collection Plan. AZA institutions are encouraged to consult the relevant Regional Collection Plan before selecting species for a conservation effort. In this way, institutions are less likely to initiate a cooperative conservation program for a species designated to be phased out of AZA collections.

Species Survival Plans©

There are currently 20 AZA primate Species Survival Plans© (SSP) covering 24 species (Table 2). There are two primary components to an SSP: (1) population management and husbandry and (2) conservation action [Wiese et al., 1994]. The SSP guides the management of species, subspecies or populations by managing the births, deaths, preferable parentage, and husbandry of the captive population. Zoo and aquarium professionals design management and breeding recommendations intended to minimize adaptation to the captive environment (i.e., domestication) and maximize the retention of gene diversity by integrating knowledge about the species' biology with studbook records. While the original efforts of the SSP focused on genetic and demographic management of the captive population, they have grown into multidisciplinary, holistic programs affecting true conservation goals. SSP Management Groups now develop and coordinate education, research, fund raising, and field conservation projects related to their species of interest [Wiese & Hutchins, 1994; Hutchins & Conway, 1995]. Several of these diverse SSP projects are documented in this volume. Participation in the SSP is voluntary, but virtually all AZA institutions holding a particular SSP species participate. SSP participation is also available to non-AZA institutions and private individuals that wish to participate once their facilities and operation have been approved by the AZA Wildlife Conservation and Management Committee.

Fauna Interest Groups

Fauna Interest Groups (FIGs) were established to coordinate and facilitate the cooperative conservation activities of AZA member institutions working in specific geographic regions of the world [Hutchins & Conway, 1995]. The AZA currently has active FIGs for Brazil, East Africa, Madagascar, Meso-America, North America, Paraguay, Southeast Asia, Venezuela, West Indies, and Zaire. Significant primate populations exist in most of the existing FIG regions. With FIG assistance, in-country conservation programs conducted by SSPs and TAGs can coordinate their activities to reduce duplication of effort as they develop cooperative zoo-based field research and habitat protection plans. Of particular importance is the development of long-term working relationships with appropriate government wildlife agencies and local non-governmental organizations and universities. This is best accomplished through an organized, cooperative approach, rather than having each institution or AZA committee contact these agencies separately. Typically, the government agencies are in the best position to identify and prioritize projects within their jurisdiction. The FIGs provide specific opportunities for member institutions to get involved in field conservation. For example, the AZA Madagascar FIG works closely with appropriate government agencies and has been very active in lemur conservation, planning for a reintroduction of ring-

TABLE 2
PRIMATE SPECIES MANAGED BY
AZA SPECIES SURVIVAL PLANS©

PROSIMIANS (6 programs covering 6 species)
Lemur, Black (*Eulemur macaco macaco*)
Lemur, Black-and-White Ruffed (*Varecia variegata variegata*)
Lemur, Coquerel's Mouse (*Mirza coquereli*)
Lemur, Mongoose (*Eulemur mongoz*)
Lemur, Ring-tailed (*Lemur catta*)
Loris, Pygmy Slow (*Nycticebus pygmaeus*)

NEW WORLD MONKEYS (3 programs covering 3 species)
Monkey, Goeldi's (*Callimico goeldii*)
Tamarin, Cotton-top (*Saguinus oedipus*)
Tamarin, Golden Lion (*Leontopithecus rosalia rosalia*)

OLD WORLD MONKEYS (6 programs covering 8 species)
Colobus, Black and White (*Colobus guereza*)
Drill (*Mandrillus leucophaeus*)
Langur, Francois' (*Presbytis francoisi*)
Macaque, Lion-tailed (*Macaca silenus*)
Mangabey, Red-capped (*Cercocebus torquatus*)
Mangabey, Golden-bellied (*Cercocebus galeritus chrysogaster*)
Mangabey, Black (*Cercocebus aterrimus*)
Monkey, De Brazza's (*Cercopithecus neglectus*)

GIBBONS (1 program covering 3 species)
Gibbon, Siamang (*Hylobates syndactylus*)
Gibbon, White-cheeked (*Hylobates concolor*)
Gibbon, White-handed (*Hylobates lar*)

GREAT APES (4 programs covering 4 species)
Bonobo (*Pan paniscus*)
Chimpanzee (*Pan troglodytes*)
Gorilla, Lowland (*Gorilla gorilla gorilla*)
Orangutan (*Pongo pygmaeus*)

tailed and ruffed lemurs and providing direct support for Parc Tsimbazaza and Parc Ivoloina [Zeeve, 1994; Zeeve & Porton, this volume].

Scientific Advisory Groups

Scientific Advisory Groups (SAGs) are discipline-oriented committees that provide assistance to AZA cooperative programs in developing research projects and solving challenges in captive propagation, reintroduction, and field conservation [Hutchins & Conway, 1995; Hutchins et al., 1996a]. AZA SAGs have been formed to provide assistance to our cooperative programs in the fields of nutrition, veterinary science, contraception, small population management, reintroduction, systematics, and behavior and husbandry. The SAG membership comes from a wide array of home institutions including the zoo and aquarium community, universities, state and federal government, and public corporations. Their efforts continue to improve both the management of captive animals and the preservation of wild populations. For example, the theoretical research conducted by members of the small population management group for captive populations is also relevant to the increasingly fragmented and fenced wild populations.

The integrated, multidisciplinary approach of the AZA C&S committee system allows the network to initiate conservation action in four broad areas: (1) public education; (2) basic and applied research (including the development of relevant technologies); (3) preservation of genetic variation in the event that future reintroduction is appropriate and necessary; and (4) direct support for field conservation [Hutchins et al., 1995; Hutchins & Conway, 1995].

PUBLIC EDUCATION

The importance of public education by modern, professionally-managed zoos cannot be over-emphasized [see Whitehead, 1995]. Public education, both within North America and abroad, is extremely important to the protection and maintenance of habitat now as well as to the success of endangered species recovery efforts. One of the advantages of working with primates is that they already enjoy great public appeal. More than 120 million people visit AZA-accredited institutions annually, and the potential for public outreach on behalf of primates and primate conservation is virtually unlimited. Education programs are quickly becoming a primary focus of Species Survival Plans© and Taxon Advisory Groups in addition to their traditional endeavors in population management.

The Fort Wayne Children's Zoo, in Indiana, has developed an interactive education and conservation action program to assist primate conservation on the Mentawi Islands, Indonesia. Through an outreach program entitled "Save the Primates," school children in Indiana are exposed to the culture of the Mentawi people and informed about the region's five endemic primate species [Priapato, 1990]. Following this exposure, the students are encouraged to take an active role in conservation. Through their classes they can exchange letters with Indonesian students and raise funds to support the work of Dr. Richard Tenaza. With the help of various zoos and conservation organizations, Tenaza has been successful in several endeavors in Indonesia. A letter writing campaign generated by Tenaza helped halt the conversion of Siberut rainforest to palm oil plantations. His work also led to progress in setting aside land throughout the Mentawi Islands for wildlife preserves [Tenaza, 1989, 1991].

Education programs in the focus species' country of origin are extremely important to the success of reintroduction projects [Durrell & Mallinson, 1987]. If attitudes of local people cannot be changed and an appreciation of wildlife fostered, then reintroduction and other conservation efforts are unlikely to be successful. Such education efforts should begin early in the recovery effort, well before actual reintroductions are attempted. When individuals are educated and asked to change their lifestyle they must also be given acceptable alternatives. If a species' decline is primarily due to hunting for food, a system of regulated sustainable harvest or alternative protein sources must be identified. If the forests are being cut for charcoal or firewood, alternative fuels must be made available. Local people must also be made aware of potential benefits surrounding the species they are asked to conserve. In some cases, ecotourism may be an option; this is especially true for primate species, which attract great interest by the general public. Three excellent examples of the application of zoo-based education and ecotourism to conservation are Kleiman [1986] for the golden lion tamarin in Brazil [and see Stoinski et al., this volume], Savage [1989, 1993, 1996, this volume] for the cotton-top tamarin in Columbia, and Horwich et al. [1993] and Horwich [1995] for the black howling monkey in Belize.

To help facilitate the use of education as a conservation tool, the AZA C&S network works closely with the AZA's Education Department and Conservation Education Committee to identify education advisors for SSPs, TAGs, FIGs and SAGs [Hutchins, 1996]. It is hoped that a full integration of zoo-based educators into the network will result in a growing number of educational initiatives, both within and outside North America.

BASIC AND APPLIED RESEARCH

Zoos are active in both basic and applied research on various primate species [Bowdoin et al., 1994; Swaringen, et al., 1995; Hutchins et al., 1996a]. Obtaining accurate knowledge of a species' basic biology is a critical first step if captive breeding efforts are to be successful in maintaining a population long term. Once information on a species' basic biology is available, applied studies related to improving animal health, controlling reproduction, and developing reintroduction and field conservation techniques are more likely to succeed. Zoos and aquariums support research on both wild and captive populations. These programs can assist broader conservation efforts through:

(1) Development of successful husbandry protocols for a variety of species (documentation of normal physiological, nutritional, reproductive, and behavioral parameters is particularly useful in this regard);

(2) Clarification of pedigrees and resolution of taxonomic status (specific and subspecific) through detailed genetic analysis;

(3) Development of veterinary procedures that may help both wild and captive populations;

(4) Development of capture, restraint, identification, monitoring, and other techniques for use in field conservation and in captive breeding programs; and

(5) Generation of new knowledge about the basic biology of various species.

Some examples of zoo-based basic research on primates include Wasser and Norton's [1993] work on sex ratio adjustment among yellow baboons in Mikumi National Park, Tanzania, and McDaniel et al.'s [1993] work on the reproductive and social dynamics of captive black-handed spider monkeys. Examples of applied research include Horwich et al.'s [1993] testing of methods to translocate and reintroduce black howling monkeys in Belize and Beck et al.'s [1987] and Kleiman's [1989] work on preparing captive-bred primates for reintroduction.

CAPTIVE BREEDING FOR REINTRODUCTION AND POPULATION MANAGEMENT

The activities of humans are pushing many primate species to the brink of extinction. The destruction of forests, due to logging and conversion to crops, has been particularly devastating [Harcourt & Thornback, 1990; Lee et al., 1988]. While captive breeding for reintroduction is not a panacea for the endangered species problem, it is still the only option for species whose populations have become so small or fragmented that they cannot survive without human intervention [Hutchins et al., 1995; Hutchins et al., 1996b]. The golden lion tamarin, of Brazil, is an excellent example of an endangered primate species that has benefited from a well-managed, scientifically-based reintroduction program [Kleiman, 1989; Stoinski et al., this volume].

Efficient management of captive primate populations is important to ensure healthy, viable populations and safeguard their long-term viability. Proper genetic and demographic management is just as critical as proper diets and veterinary care [Wiese et al., 1994]. Without proper genetic management, individual animals will experience increased health problems in the short-term and the populations' ability to survive in the long-term may be reduced. Genetic management, therefore, is critical for any population for which reintroduction is contemplated. Maintaining sustainable captive populations also helps reduce the need to collect animals from the wild, lowers the cost of animal acquisition, and reduces the economic incentive for the illegal wildlife trade [Hutchins et al., 1996b].

The primary goal of collective population management by AZA member zoos is to retain as much gene diversity as possible, thereby minimizing genetic adaptation to the captive environment. Retention of gene diversity is accomplished by careful monitoring of which individuals are allowed to breed. Based on pedigree analysis, animals with the rarest alleles receive preference for breeding over those with more common alleles found in the population. Individuals with rare alleles are allowed to breed more often, thereby producing more offspring and increasing the frequency of the rare alleles. Animals are paired with other individuals with similar allele frequencies in the population to prevent combining rare and common alleles in the offspring. Combining rare and common alleles in individual offspring prevents future equalization of allele frequencies through differential breeding of founder lines. In North America, selection of breeding pairs is determined by examining an individual's mean kinship value which is a measure of how related an individual is to the population [Ballou & Lacy, 1995]. Individuals with lower mean kinship values (i.e., fewer relatives) possess the rarest alleles in the population. Space in zoos and aquariums is limited and, in many cases, the SSP must limit the number of births to ensure the best housing conditions for the animals and to avoid overcrowding.

Retention of gene diversity allows the population to minimize inbreeding in the long-term which decreases the expression of deleterious alleles. This assures better health of individual animals and reduces concern over some animal welfare issues [Wiese et al., 1994]. SSPs also develop and distribute husbandry manuals that increase the understanding and consistency of care for captive populations across holding institutions.

While genetic management is a critical component of responsible animal management, it must be accomplished in ways that complement the behavioral biology of a species. This is particularly important for social animals like primates. In many cases the difficulty of moving animals into or out of stable social groups precludes perfect genetic management based on theoretical models. For example, in some primates that live in large social groups in nature (e.g., macaques), adult females form the stable core and the tenure of adult males can be relatively brief. In such cases, genetic management in captivity can be accomplished by rotating males between groups, while maintaining the female social structure. Natural dispersal can also be simulated by moving young animals between groups.

SUPPORT OF FIELD CONSERVATION

In recent years direct support of field conservation initiatives has received increased focus by the zoo and aquarium community [Hutchins & Wiese, 1991; Hutchins & Conway, 1995]. Because this aspect of zoo-based primate conservation is detailed elsewhere in this volume, we will provide only a few brief comments here.

Considering the number of projects currently underway, it is obvious that the AZA and its members consider primates a high priority. Zoo-based contributions to primate conservation in nature now range from applied field research to habitat protection to assistance with conservation planning. Examples of the former are given throughout this chapter and by Koontz [this volume]. An example of assistance with conservation planning is provided by Schaaf's [1990] cooperative work with the Republic of Equatorial Guinea. Schaaf and his colleagues from Zoo Atlanta assisted government officials by conducting a survey of drill populations in the Gran Caldera Volcanica de Luba and in conservation planning for this endangered primate [Schaaf et al., 1990]. Similarly, the Dian Fossey Gorilla Fund, which focuses its attention on the conservation of free-ranging mountain gorillas, is now based at Zoo Atlanta [Anonymous, 1996].

CONCLUSIONS

The AZA Conservation and Science network is organized to facilitate cooperative conservation action by the AZA and its 180 member institutions. AZA institutions currently maintain 53 studbooks on various primate species, 5 primate advisory groups, and 20 primate Species Survival Plans© covering 24 species. In addition, cooperative field conservation and scientific initiatives are organized through Fauna Interest Groups and Scientific Advisory Groups, respectively. This places North American zoos in a strong position to become major players in primate research and conservation worldwide, both individually and collectively through their parent organization, the AZA. The AZA and the American Society of Primatologists recently established an official liaison to help improve communication and collaboration.

There is much work to be done if we are to assist even a fraction of the primate species currently at risk. However, through the successful use of captive breeding and reintroduction, scientific research, *in situ* conservation efforts and public education, professionally managed zoos can play a significant role in primate conservation.

ACKNOWLEDGMENTS

We thank K. Willis for reading and commenting on the manuscript, J. Wallis for organizing the symposium at which this paper was presented, and the American Society of Primatologists for their interest and support of zoo-based primate conservation and science.

REFERENCES

Anonymous. Dian Fossey Gorilla Fund relocates to Zoo Atlanta from Denver, Colorado. ENDANGERED SPECIES UPDATE 13:15, 1996.

Ballou, J.D.; Lacy, R.C. Identifying genetically important individuals for management of genetic variation in pedigreed populations. Pp. 76-111 in POPULATION MANAGEMENT FOR SURVIVAL AND RECOVERY. J.D. Ballou; M. Gilpin; T.J. Foose, eds. New York, Columbia University Press, 1995.

Beck, B.B.; Castro, I.; Kleiman, D.; Dietz, J.M.; Rettberg-Beck, B. Preparing captive-born primates for reintroduction. INTERNATIONAL JOURNAL OF PRIMATOLOGY 8:426, 1987.

Bowdoin, J.; Wiese, R.J.; Willis, K.; Hutchins, M., eds. AZA ANNUAL REPORT ON CONSERVATION AND SCIENCE. Bethesda, Maryland, American Zoo and Aquarium Association, 1994.

Durrell, G.; Mallinson, J. Reintroduction as a political and educational tool for conservation. DODO 24:6-19, 1987.

Harcourt, C.; Thornback, J. LEMURS OF MADAGASCAR AND THE COMOROS. The IUCN Red Data Book. Gland, Switzerland, IUCN, 1990.

Horwich, R.H. Community-based ecotourism in Belize, Central America. Pp. 243-249 in AZA ANNUAL CONFERENCE PROCEEDINGS. Wheeling, West Virginia, American Zoo and Aquarium Association, 1995.

Horwich, R.H.; Koontz, F.; Saqui, E.; Saqui, H.; Glander, K. A reintroduction program for the conservation of the black howler monkey in Belize. ENDANGERED SPECIES UPDATE 10:1-6, 1993.

Hutchins, M. Education is conservation. AZA COMMUNIQUÉ March:26-27, 1996.

Hutchins, M.; Conway, W. Beyond Noah's Ark: The evolving role of modern zoological parks and aquariums in field conservation. INTERNATIONAL ZOO YEARBOOK 34:117-130, 1995.

Hutchins, M.; Paul, E.; Bowdoin, J. Contributions of zoo and aquarium research to wildlife conservation and science. Pp. 23-39 in THE WELL-BEING OF ANIMALS IN ZOO AND AQUARIUM SPONSORED RESEARCH. G. Burghardt; J. Bielitzki; J. Boyce; D. Schaeffer, eds. Greenbelt, Maryland, Scientist's Center for Animal Welfare, 1996a.

Hutchins, M.; Wiese, R. J. Beyond genetic and demographic management: The future of the Species Survival Plan and related AAZPA conservation efforts. ZOO BIOLOGY 10:285-292, 1991.

Hutchins, M.; Wiese, R.J.; Willis, K. Why we need captive breeding. Pp. 77-86 in AZA REGIONAL CONFERENCE PROCEEDINGS. Wheeling, West Virginia, American Zoo and Aquarium Association, 1996b.

Hutchins, M.; Wiese, R.J.; Willis, K.; Becker, S., eds. AAZPA ANNUAL REPORT ON CONSERVATION AND SCIENCE. Bethesda, Maryland, American Association of Zoological Parks and Aquariums, 1991.

Hutchins, M.; Willis, K.; Wiese, R.J. Strategic collection planning: Theory and practice. ZOO BIOLOGY 14(1):2-22, 1995.

Kleiman, D.G. Conservation program for the golden lion tamarin: captive research and management, ecological studies, educational strategies and reintroduction. Pp. 959-979 in PRIMATES: THE ROAD TO SELF-SUSTAINING POPULATIONS. K. Benirschke, ed. New York, Springer-Verlag, 1986.

Kleiman, D.G. Reintroduction of captive mammals for conservation. BIOSCIENCE 39:152-161, 1989.

Lee, P.C.; Thornback, J.; Bennett, E.L. THREATENED PRIMATES OF AFRICA. THE IUCN RED DATA BOOK. Gland, Switzerland, and Cambridge, UK, IUCN, 1988.

McDaniel, P.S.; Janzow, F.T.; Porton, I.; Asa, C.S. The reproductive and social dynamics of captive *Ateles geoffroyi* (Black-handed spider monkey). AMERICAN ZOOLOGIST 33:173-179, 1993.

Priapato, C. Saving the primates: Fort Wayne Children's Zoo heads international effort. ZOOTAILS (Fort Wayne Children's Zoo) November/December: F1-F4, 1990.

Savage, A. Proyecto titi: A hands on approach to conservation education in Colombia. Pp. 605-606 in AAZPA ANNUAL CONFERENCE PROCEEDINGS. Wheeling, West Virginia, American Association of Zoological Parks and Aquariums, 1989.

Savage, A. Tamarins, teens and teamwork: An integrated approach to *in situ* conservation. Pp. 106-108 in AAZPA ANNUAL CONFERENCE PROCEEDINGS. Wheeling, West Virginia, American Association of Zoological Parks and Aquariums, 1993.

Savage, A. AZA Species Survival Plan profile: The cotton-top tamarin. ENDANGERED SPECIES UPDATE 13:9-11, 1996.

Schaaf, D. Search for the drill. ZOOMAGAZINE (Zoo Atlanta) November/December:2-3, 12. 1990.

Schaaf, D.; Butynski, T.M.; Hearn, G. The drill and other primates in the Gran Caldera Volcanica de Luba: Results of a survey conducted March 7-22, 1990. Unpublished report to the government of the Republic of Equatorial Guinea. 1990.

Swaringen, K.; Wiese, R.J.; Willis, K.; Hutchins, M., eds. AZA ANNUAL REPORT ON CONSERVATION AND SCIENCE. Bethesda, Maryland, American Zoo and Aquarium Association, 1995.

Tenaza, R.R. Primates on a precarious limb. ANIMAL KINGDOM 92:27-37, 1989.

Tenaza, R.R. The Mentawi Islands: A case study in conservation of rainforest primates and archaic culture. Pp. 602-606 in AAZPA REGIONAL CONFERENCE PROCEEDINGS. Wheeling, West Virginia, American Association of Zoological Parks and Aquariums, 1991.

Wasser, S.K.; Norton, G. Baboons adjust secondary sex ratio in response to predictors of sex-specific offspring survival. BEHAVIORAL ECOLOGY AND SOCIOBIOLOGY 32:273-281, 1993.

Whitehead, M. Saying it with genes, species and habitats: biodiversity education and the role of zoos. BIODIVERSITY AND CONSERVATION 4:664-670, 1995.

Wiese, R.J.; Hutchins, M. SPECIES SURVIVAL PLANS: STRATEGIES FOR WILDLIFE CONSERVATION. Bethesda, Maryland, American Zoo and Aquarium Association, 1994.

Wiese, R.J.; Hutchins, M.; Willis, K.; Becker, S., eds. AAZPA ANNUAL REPORT ON CONSERVATION AND SCIENCE. Bethesda, Maryland, American Association of Zoological Parks and Aquariums, 1992.

Wiese, R.J.; Willis, K.; Bowdoin, J.; Hutchins, M., eds. AAZPA ANNUAL REPORT ON CONSERVATION AND SCIENCE. Bethesda, Maryland, American Association of Zoological Parks and Aquariums, 1993.

Wiese, R.J.; Willis, K.; Hutchins, M. Is genetic and demographic management conservation? ZOO BIOLOGY 13:297-299, 1994.

Zeeve, S. A FIG turns six: The Madagascar Fauna Group in 1994. Pp. 141-145 in AZA ANNUAL CONFERENCE PROCEEDINGS. Wheeling, West Virginia, American Zoo and Aquarium Association, 1994.

Dr. Robert J. Wiese is currently the Assistant Director of Animal Collections at the Fort Worth Zoological Park. Formerly, he was Assistant Director, Conservation and Science for the American Zoo and Aquarium Association. A geneticist by training, Dr. Wiese is chair of the AZA's Small Population Management Advisory Group. **Dr. Michael Hutchins** is AZA Director of Conservation and Science and Adjunct Professor at the University of Maryland's Graduate Program in Sustainable Development and Conservation Biology. He is also an Associate Editor of *Zoo Biology* and primary Editor of the Smithsonian Institute Press's book series: *Studies in Zoo and Aquarium Biology and Conservation.*

A red uakari (*Cacajao calvus*) is viewed by visitors along the South American rain forest at the Monkey Jungle, circa 1965. Visitors entered the animals' environment, becoming immersed in a simulated rain forest. (Photo courtesy of Sharon Dummont, Monkey Jungle. Photographer unknown.)

THE CONSERVATION ROLE OF PRIMATE EXHIBITS IN THE ZOO

Kenneth C. Gold
Lincoln Park Zoo, Chicago, Illinois

INTRODUCTION

Zoological parks have the ability to educate hundreds of millions of people annually about wildlife and nature and influence their attitudes toward conservation. Approximately 600 million people visit zoos annually worldwide; in the United States alone, more than 117 million people visit zoos each year to view exotic animals [Wagner, 1996].

Because live animals have a natural ability to attract the public's attention, the zoo is an ideal place to educate people about the world's flora and fauna. For many people, the zoo is the first (and possibly only) place they will encounter any of these rare creatures. Often, particularly in urban areas, a person can grow up with no natural contact with nature. Zoos provide the setting in which to learn the reality of natural habitats, ecological and biological principles, and the impact of humans on plants and animals.

The way in which zoo animals are displayed can have a profound influence on the public's attitude toward that animal (individual or species), and what they learn about its natural history, lifestyle, role in its natural environment and conservation status [Coe & Maple, 1984; Ogden et al., 1991]. Zoos have become conservation educators with unique opportunities (and responsibilities) to help people understand the interconnection of all living things. The concepts of biodiversity, biological interdependency and personal responsibility are major messages incorporated into a modern zoo's mission. However, this was not always the case.

The exhibition of zoo animals has changed dramatically over the past 50 years, with conservation becoming a key issue primarily during the last two decades. Zoo exhibit design has evolved throughout the years from barren cages or pits to hygienic tile-walled, glass-fronted "bathroom-style" enclosures to naturalistic museum-type dioramas to the recent naturalistic exhibits with moated, glass or mesh barriers [Gold, 1992]. This evolution in exhibitry has been driven by changes in the public's attitude and an evolving understanding of the physical and social needs of nonhuman animals.

Primates are some of the most popular animals in the zoo. Great apes are of special interest, primarily due to their anatomical and behavioral similarities to humans, high level of intelligence, and natural curiosity [Gold & Benveniste, 1995]. Zoo exhibit design has played a role in primate conservation in three main ways. First, exposing the visitor to these rare and unique creatures in an accessible setting facilitates the education of visitors about a multitude of concepts such as animal behavior, status of their natural environment, and ecosystem preservation. Second, by designing exhibits that stimulate captive breeding, zoos can ease the pressure on the remaining wild populations, eliminating at least one of the markets for trade in primates. In a few cases, such as with the golden lion tamarin [Stoinski et al., this volume], breeding programs may facilitate the successful reintroduction of captive born animals into the wild. Third, a few zoos have encouraged their visitors to take action, to become conservationists as a result of viewing zoo exhibits. This chapter will trace the evolution of primate zoo exhibit design and the role conservation has played in this evolution.

SCIENCE AND CONSERVATION GOALS OF EARLY ZOOS

Several early zoos, including the London Zoo, the National Zoo (Smithsonian Institution) and the Bronx Zoo (New York Zoological Society), had both education and scientific inquiry as goals. In 1828, the Zoological Society of London opened the zoological garden in Regent's Park with the stated purpose of studying captive animals to better understand their wild relatives. This served as a model for future zoological gardens throughout Europe and America. In the 1870s, a menagerie was created as part of the Smithsonian Institution, Washington, DC, primarily to provide Smithsonian taxidermists living models to aid their preservation of museum specimens. The menagerie became the National Zoo in 1891, when it opened its doors with a mission to study animals while exhibiting them for the public.

In 1895, the New York Zoological Society was formed with conservation as its main aim. The stated purpose of the institution was "...encouraging and advancing the study of zoology...furnishing instruction and recreation to the people" [Goddard, 1995, p. 42]. The Society opened the Bronx Zoo in 1899, designed as a wildlife preserve "...in which living creatures can be kept under conditions most closely approximating those with which nature usually surrounds them" [p. 42]. The park was intended to be a place of study and inspiration, as well as recreation, where the Society could "...cultivate in every possible manner the knowledge and love of nature" [p. 42]. While these ambitious goals of early zoos were certainly praiseworthy, their implementation, especially regarding animal exhibitry, was a slow process.

THE EVOLUTION OF ZOO PRIMATE EXHIBITS

The First Exhibits

Traditionally, zoos exhibited primates in cages with barren surroundings. Little was known of the animals' natural environment or behavior, and this type of exhibit design was created primarily to allow the visitor a close and unobstructed view of these unique creatures — while still maintaining visitor safety. There was little attention given to the animals' needs; animal welfare was of secondary concern. Exotic animals were viewed as expendable curiosities; if one died, it could easily be replaced with another from the wild [Gold, 1992].

Some exhibits were designed to reflect the animals' geographic origin. This "atmospheric" style created an elaborate, picturesque effect rather than catering to the physiological needs of the animals. In addition, the pagodas, temples and mosques used in the exhibits failed to take into account the cultural, scientific and instructional role that zoos could promote [van den Bergh, 1960]. Other exhibits were designed to highlight the physical attributes of the inhabitants. Some of these exhibits tended to be overbuilt, emphasizing the dangerous nature of the animal. This was especially true for some of the apes, whose supposed 'ferociousness' was used in marketing strategies to attract visitors. Visitors were intrigued by captive primates, which became favorites of the zoo-going public. Many early exhibits were designed around a taxonomic theme, e.g., groupings of many primate species together in a primate house. Some zoos, such as Lincoln Park Zoo in Chicago, still exhibit their animal collection taxonomically.

Whereas the primary intention of these early displays was to bring visitors to see the rare and exotic animals, occasionally some of these exhibits educated the public about the animals' natural history or conservation status. Antwerp Zoo, founded in 1843, was one of the first zoos to include scientific labels, incorporating colored pictures of the animals, a distribution map, and a short description of the species' natural habitat.

In 1896, explorer Robert Garner proposed several visionary ideas for housing great apes based on observations of their physiology and behavior in the wild. He recommended that all apes should have enclosures at least 15 feet in height, an earthen floor, and a wide but shallow pool. He suggested that the temperature should be allowed to vary between conditions normal for humans, and the exhibit should include an intermittent artificial rain spray. He also recommended inclusion of a strong tree for exercise and piles of dead leaves provided for the animals' comfort [in Hancocks, 1971].

Many of these ideas were incorporated into the design of primate houses around the turn of the century. William T. Hornaday [1913], then Director of the Bronx Zoo, described the design of their primate house, which opened in 1901: "Points of special excellence in this building are as follows: An abundance of room for the animals, an abundance of sunlight, perfect ventilation, an extensive series of open-air cages, freedom of communication between outside and inside cages, floors and walls impervious to moisture and disease germs, and the absence of iron bars from all cages save three" [pp. 80-81]. Japanese macaques and certain species of baboons were exhibited outdoors year-round. To educate visitors about the taxonomy of primates, the Bronx Zoo exhibited their animals in 4 groupings: apes (chimpanzee, orangutan and gibbon), Old World monkeys, New World monkeys, and lemurs.

In 1907, the London Zoo described their new primate house as one of the first mixed species primate exhibits created [Loisel, 1908]. The ape house contained four large cages separated from the public corridor by a glass plate. This design, one of the first to incorporate glass as a barrier, shielded apes from public contact, prevented public feeding and preserved a uniform temperature in the enclosure.

After viewing many of the primitive, barren conditions in which captive primates often failed to thrive or even perished, exhibitors (and the public) began to question if there were better ways to house animals. Seton [1901] commented that the higher apes and baboons rarely thrive in cages: "Sooner or later they become abnormally vicious, or else have a complete physical breakdown. Occupation and plenty of good food are not the only things needful for a well-rounded life" [p. 709]. Most zoos of the time were keeping monkeys in "stuffy hot-houses," which often did nothing to prevent the constant respiratory maladies of the inhabitants. Seton reported that one European zoo ceased "coddling" primates, instead giving them free access to fresh air and sunlight and a cage large enough to permit exercise. As a result, the illnesses disappeared. He also noted that, due to the extremely unsuccessful keeping of captive gorillas, the authorities discouraged dealers from importing these apes.

Barless Enclosures

In 1906, Carl Hagenbeck revolutionized the zoo world when he designed exhibits for animals in "natural surroundings" and pioneered the use of a moat as a barrier across which one could view animals. His Tierpark, near Hamburg, Germany, eliminated barred enclosures and included natural elements – such as rocky cliffs for baboons. The concept of removing the bars and creating moated exhibits for primates had a profound influence, leading to a proliferation of "monkey islands" in zoos worldwide. They are the forbears of many of the modern naturalistic primate exhibits of today. Hagenbeck also built exhibits that included replicas of architectural elements of the animals' homelands (such as Mayan temples) and, unfortunately, some zoos even imported indigenous people to display alongside the animals. Bristol Zoo's first barless enclosure, the Monkey Temple, was built in 1929, and displayed rhesus macaques in an Indian temple. Some of these cultural architectural elements are making a comeback in modern zoos to connect visitors to the primate species' exotic homelands.

Throughout the 1920s and 1930s many zoos created monkey islands in the Hagenbeck tradition, primarily for baboons and macaques. The first such exhibit in the U.S. opened in 1929, when Baboon Rock was completed at the Detroit Zoo (Figure 1). In 1936, Detroit created another monkey island and stocked it with

Figure 1. Baboon Rock at the Detroit Zoo, circa 1929. The huge crowd attests to the popularity of this exhibit. (Photo courtesy of Ron Kagan, Detroit Zoo. Photographer unknown.)

200 rhesus monkeys. Blair [1937] stated that "outdoor exhibits of large numbers of the primates, in barless enclosures, are a comparatively recent development in zoological park planning, but there is not one installation ancient or modern that exceeds them in permanent popular interest" [p. 71]. Due to the popularity of the Bronx Zoo's primate house, a monkey island was planned as a practical method of exhibiting large colonies of monkeys, "...under such conditions that they can be seen distinctly and comfortably viewed by thousands of visitors at a time" [p. 71].

The first ape exhibit using water as a barrier appeared in 1953 with the Dublin Zoo's gibbon islands. In 1956, an outdoor moated island for chimpanzees was constructed at Chester Zoo, the first for great apes [Mottershead, 1960]. The main idea at Chester was to permit the chimpanzees as much time outdoors as possible. Multiple islands, surrounded by shallow water moats 3 feet deep at the deepest point and 17 feet wide, were connected via tunnels to their indoor exhibits. This exhibit demonstrated to the zoo world the ability to contain apes on outdoor islands with natural substrates, another major step in the evolution of zoo exhibit design for great apes. Thus, several large moated exhibits for apes were developed in the mid 1960s, notably at the Los Angeles Zoo and Busch Gardens in Tampa, Florida.

Figure 2. Long-tailed macaques (*Macaca fascicularis*) at Monkey Jungle in Goulds, Florida. More than 80 monkeys live in this rain forest exhibit, the first of its kind in North America. (Photo by Ken Gold.)

The First Naturalistic Exhibits

In 1935, a spectacular experiment in exhibition began when Monkey Jungle opened a 5 acre hammock forest in sub-tropical south Florida. It originally housed a small group of long-tailed macaques, which has since grown to approximately 80 animals (Figure 2). In the late 1950s, a second free-ranging 4 acre habitat was opened at Monkey Jungle known as the "rainforest." This area was modified with the addition of over 180 species of tropical American forest trees, creating a lush environment for the nonhuman primate inhabitants. Today it includes more than 100 squirrel monkeys, red howlers, black-capped capuchins and Goeldi's monkeys. Historically, it has also housed groups of red uakaris, saddle-backed tamarins, titi monkeys, white-faced saki monkeys, golden lion tamarins and cotton-top tamarins [Evans et al., in press] (see photo on face page of chapter). This was one of the first examples of a zoogeographic theme – with the animals being arranged according to their continent of origin, combined with a bioclimatic theme – with the animals arranged according to habitat (in this case rainforest).

Both of these forests within Monkey Jungle provide habitat for the semi-free-ranging primates, while visitors are contained along a series of fenced-in trails. Initially, the rainforest habitat trails were not fenced-in, but the aggressiveness of the uakari group necessitated the eventual fencing for visitor safety. According to Frank DuMond [1968], the rainforest habitat was created as a "living museum" for the public. Although education about conservation was limited at Monkey Jungle, even the casual visitor learned much about primates, their behavior and their relationship to the environment. In addition, many scientific studies have been conducted at Monkey Jungle, including research on reproduction, ontogenic and social behavior, locomotor behavior, and patterns of learning and sensory skills of a number of species.

The Hygienic Phase of Exhibit Design

During the 1950s another major trend, viewed by many as regressive, appeared in the housing and exhibition of primates – especially apes. At the time, many animals were succumbing to early death due to diseases. The fear of disease transmission led to the return of sterile indoor habitats for primates; zoos designed clinical habitats - made of tile, glass and stainless steel. They were easy to clean and disinfect, thereby reducing the chances for the spread of disease. Lang [1960] wrote that modern methods of keeping apes are based on the knowledge that they are extremely susceptible to colds and human diseases that can lead to dangerous or fatal illnesses. He stated that they "...must, above all else, be protected from infections carried by human beings. A glass screen erected in front of both indoor and outdoor cages effectively excludes these elements of danger" [p. 3].

van den Bergh [1959] described the design of the new ape house at Antwerp Zoo. The walls of the cage were covered with enameled concrete, the floors were tiled, the exhibit fronts were hermetically sealed glass, and a solid iron trellis protected a glass ceiling. Most of the cages were fitted with metal pipe furnishings for exercise and play, and one included a weight scale with a read-out dial on the visitor side of the display. The public gallery was in front of the day cages and was very spacious and well ventilated. The cages were brightly lit and the public gallery was kept in darkness. The cages were equipped with microphones connected to loudspeakers in the visitor areas allowing broadcast of vocalizations and other sounds made by the animals. Instructional showcases were set into the wall, containing photographs, anatomical specimens and graphic charts giving details of ape biology. These interpretive methods have been emulated in many zoos worldwide.

Whereas captive environments for primates have improved dramatically over the last 20 years, there are still many that lack stimulation and ignore many of the needs of the inhabitants. These exhibits often promote atypical behaviors as a result of the inappropriateness of exhibit design, which generally fail to take into account the natural adaptations of primates [Coe & Maple, 1984; Gold, 1992]. For example, concrete or tiled boxes with a few fixed metal structures correspond poorly to the complex environment in which apes live in the wild [Mallinson et al., 1994]. Fortunately, through the use of behavioral or environmental enrichment,

Figure 3. A silverback male gorilla (*Gorilla g. gorilla*) in the lushly-planted rain forest exhbit at Seattle's Woodland Park Zoo. This was the first great ape exhibit to immerse visitors in the same landscape atmosphere as the animals, eliminating visible barriers. (Photo by Ken Gold.)

many institutions at least supplement these old-style exhibits to stimulate the natural behavior of the animals.

Nocturnal habitats were developed by zoos in the late 1950s and early 1960s. Early nocturnal houses in London, New York, Pittsburgh, Antwerp and Amsterdam displayed bush babies, owl monkeys, lorises, and dwarf lemurs. The Detroit Zoo, in 1962, was one of the first North American zoos to experiment with red lights, which permitted viewing of nocturnal primates under a reversed light cycle condition. This allowed the zoo visitors a first opportunity to view nocturnal species in their normal night-active state.

In the 1960s, zoos began to learn more about animals in their natural habitats from field researchers like Jane Goodall and George Schaller. Some of the information from this expanding accumulation of field research was incorporated into zoo exhibit design. In addition, captive studies of primate behavior were also being conducted. Combined, these studies of primate behavior led zoo planners to design more appropriate enclosures.

Immersion Exhibits

In 1975, the architectural firm of Jones and Jones, in collaboration with David Hancocks, Director of the Woodland Park Zoo in Seattle, devised an exhibit approach labeled "habitat" or "landscape immersion" [Jones et al., 1976; Process Architecture, 1995]. After extensive research of the habits and habitats of the animals in question, the design team created a realistic replica of the animals' indigenous habitat using vegetation, geological formations, state of the art technology, sight-line design, and staging. Immersion exhibits – creating the illusion that there are no barriers between visitors and animals – were championed with the completion of the Woodland Park Zoo gorilla exhibit in 1978 (Figure 3). Landscape immersion exhibits began a second revolution in the philosophy of exhibit design by recreating nature for the animals and placing the visitor in a "natural place," too.

The intention of an immersion exhibit is to provide for the visitor a feeling of awe as well as a sense of place, creating a heightened sense of awareness due to the lack of apparent physical barriers [Coe, 1985]. There is an attempt to make the visitors a little uncomfortable, as though they are intruders into the animals' world. Although not fully tested, it is thought that immersion design can affect zoo visitors stronger than traditional settings where the visitors are clearly aware of barriers [Finley et al., 1988; Shettel-Neuber, 1988]. These exhibits create an emotional predisposition to learn more about the animal. The visitors come to understand that wild animals and wild habitats are inseparable and that neither can survive alone. If the visitor appreciates the experience, he or she may care more about the animals and their fate in nature [Koebner, 1994].

Naturalistic and Functional Habitats

During the time that habitat immersion was being developed in America, the Apenheul Naturepark in Appeldoorn, Netherlands, was opened in 1976. Apenheul is a zoo within a nature preserve with the principal focus on primates. Most of the primates (with the exception of gorillas, gibbons and capuchins) are free-roaming among the visitors. There is a strong emphasis on naturalistic and functional habitats. Functional habitats are those that meet the animals' physical and psychological needs but may not necessarily be aesthetically pleasing to humans. A conservation slant is conveyed at Apenheul, with primate exhibits divided along a zoogeographic theme. Most of the areas are named for a primate conservationist who conducted research in that particular zoogeographic region, including Dian Fossey, Russell Mittermeier, Allison Jolly, Peter Scott, Bernhard Grizmek and Adelmar Coimbra-Filho.

A second type of naturalistic primate habitat design also emerged in the 1970s – that of the indoor tropical house. Particularly in regions with colder climates, indoor habitats evolved from the sterile hygienic tile-walled exhibits into naturalistic re-creations of indigenous primate habitats. These indoor exhibits developed in two directions: large multi-species indoor habitats or a series of smaller single

Figure 4. Lowland gorillas (*Gorilla g. gorilla*) at the Brookfield Zoo's Tropic World, one of the first large-scale indoor naturalistic primate exhibits. (Photo by Ken Gold.)

species museum diorama exhibits. New materials and techniques permitted the sculpting of artificial environments with realistic appearance and features such as rockwork, artificial trees, and vines.

Brookfield Zoo's Tropic World is a good example of a large multi-species exhibit. It has three sections: tropical Asia, Africa and South America. Each area is a large room with naturalistically designed (yet artificially constructed) features. The Asian section displays gibbons, orangutans, langurs, and small-clawed otters. Inside the African exhibit are three enclosures – one that displays colobus, Kolb's guenons, mandrills and sooty mangabeys (along with a pygmy hippo). On the other side of a wooden bridge are two gorilla enclosures, one extremely large with a massive mountain, waterfall and river, and many large artificial trees, vines and boulders (Figure 4). The other is viewed through a bamboo blind and houses a solitary aged female gorilla. Birds are free-flying throughout the building. A third section has a South American theme, with golden lion tamarins, capuchins, spider monkeys, squirrel monkeys, sloths, tapirs and a giant anteater. By creating this kind of exhibit experience, the zoo hopes to give visitors "...an appreciation for the precious areas of the earth, a better understanding of the rainforest and our interdependence with it, and so in the end a desire to help conserve it" [Cherfas, 1984, p. 142]. Tropic World was conceived as a way to use the zoo's space and facilities to create viable breeding groups of animals and at the same time provide for more effective education of the visitors, teaching them about conservation by offering the opportunity to immerse themselves in the exhibit [Cherfas, 1984].

Similar exhibits including more natural materials have since been created at the Bronx Zoo (Jungle World; Figure 5) and Omaha's Henry Doorly Zoo (the Lied Jungle). Some of these primate exhibits were designed with simulated rainstorms occurring periodically over the animal areas. For instance, twice a day in Tropic World's three exhibits a five minute rainstorm unleashes above the animal habitats. This has proven to be educational for visitors in addition to stimulating the activity of the animals. In this style of exhibit, visitors are immersed in the landscapes along the habitat trails, although graphics and other educational interactive displays may detract slightly from the immersion experience.

The other direction that evolved for naturalistic indoor habitats is exemplified by the Lincoln Park Zoo's Brach primate house. Here, an old style facility with 33 old tile-walled enclosures was renovated to display eight new naturalistic habitats for primates, including gibbons, lemurs, Diana monkeys, tamarins, marmosets, mandrills, colobus monkeys, howling monkeys, squirrel monkeys and titi monkeys. Two of the exhibits house two species of primates mixed together, whereas the other six exhibits display single species. While the exhibits immerse the animals in naturalistic landscapes, the visitors remain in a traditional viewing hall observing the animals from behind glass. Designers focused on the naturalistic appearance of these exhibits, providing painted murals to reflect the original habitats of the animals.

Outdoor primate habitats have continued to evolve into re-creations of actual indigenous habitats, such as the Myombe Reserve at Busch Gardens in Tampa and the Jungle Trails exhibit at the Cincinnati Zoo and Botanical Gardens. Design teams visited field sites to understand better how to realistically display captive primates. Another extension of habitat immersion is the "cultural resonance" direction that zoo architectural firms are championing, incorporating aspects of the native culture in a naturalistic way. For example, Zoo Atlanta's orangutan exhibit is modeled after an orangutan reintroduction site in Sumatra. The gelada baboon exhibit at the Bronx Zoo immerses the visitor in an Ethiopian village setting, which includes a replica of an archeological "dig."

The evolution of exhibits has taken a new approach in some locations where free-roaming primates are displayed with a strong conservation message. Both golden lion tamarins and cotton-top tamarins are allowed to roam freely in several zoos in Europe and the U.S. [Savage et al., this volume; Stoinski et al., this volume]. The Apenheul Sanctuary in the Netherlands is experimenting with free-roaming outdoor displays of several species of primates including tamarins, squirrel monkeys, barbary macaques, saki monkeys, woolly monkeys and lemurs. The free-roaming golden lion tamarin displays serve a dual purpose — as a means of providing conservation education and as a half-way house for the animals. They allow the animals to practice navigation and foraging skills before being released

Figure 5. White-cheeked gibbons (*Hylobates concolor*) in the Asian Forest of the Wildlife Conservation Park's (Bronx Zoo) Jungle World. This naturalistic facility set the standard for indoor tropical houses for primates in the 1980s, while providing a strong conservation message about the animals and plants on display. (Photo by Ken Gold.)

into their natural range in Brazil as part of the golden lion tamarin conservation program [Stoinski et al., this volume]. While these exhibits require extensive work force to monitor the animals (frequently with radio-telemetry equipment), the resulting conservation education message is powerful and memorable for the visitor.

VISITOR ATTITUDES AND EVALUATION

Wildlife documentaries have provided the public with enhanced appreciation of nonhuman primates; people are no longer content to see these animals displayed in bare cages. Therefore, habitat immersion exhibits have proven extremely popular with zoo visitors. A visitor evaluation study of the orangutan and bonobo habitats at the San Diego Zoo compared animals in older concrete moated exhibits to those in more naturalistic exhibits [Shettel-Neuber, 1988]. The results demonstrated that visitor attitudes were much more positive toward naturalistic exhibits. Similarly, Finley, et al. [1988] found that people overwhelmingly preferred naturalistic enclosures to small, barren, and sterile exhibits.

Most people visiting zoos prefer the open view enclosures, as they give the appearance that the animals are almost free-ranging [Cherfas, 1984]. However, it is the quality of the enclosure that is more important than quantity of space. Zoo visitors often perceive most exhibits as too small, projecting their own feelings about the need for space rather than considering the animals' needs. Gold and Benveniste [1995] found that despite unprecedented success in gorilla reproduction and rearing, many visitors still perceived the Lincoln Park Zoo's Great Ape House as insufficient, ignoring the functionality and complexity provided by the design.

The often misguided perception of visitors is not a new phenomenon. Hediger [1957] remarked: "It is a great pity that so few zoos make any attempt to counter uninformed criticism of the functionality of exhibit design... In most cases, if the biological needs of an animal are the prime consideration governing the design of its enclosure, this will also be the best way of showing it to the public, because it will be encouraged to behave normally" [p. 144]. Of course the ideal solution appears to be a functionally and aesthetically naturalistic design, thus serving the needs of the animal and providing the ecological reference points to allow conservation education for the visitor.

With increased attention given to the natural requirements of zoo animals comes the enhanced well-being of the animals and, in turn, more effective educational opportunities to learn about the animals' natural behaviors. Contrary to public sentiment, orangutans, gibbons and many arboreal monkey species may be better suited to mesh enclosures that provide them additional climbing opportunities (more surface area) and generally more useable vertical space. Whereas zoos try to cater to public demand, it is, in fact, the zoos that help shape public demand. To better educate a visitor about animals, the zoo must provide the animals a setting in which they can express themselves as naturally as possible.

Surveys have assessed both the effective methods of education and topics of interest to zoo visitors. This is a new area of research in zoos and has been conducted at only a few institutions via visitor interviews and tracking techniques [Birney, 1988; Gold, 1995; Gold & Benveniste, 1995; Normandia, 1986; Serrell, 1978]. Results demonstrate that visitors are interested in conservation messages such as the animals' plight in the wild, ways to avoid extinction, primate intelligence and ecology [Gold & Benveniste, 1995].

INTERPRETIVE DISPLAYS: PAST AND PRESENT

Exhibit signs and interpretive displays have improved drastically since the days when all that was provided was a simple label with the animal's common name, a map showing the place of origin, and a few words about diet, anatomy or behavior [Serrell, 1988]. Research demonstrates that people learn in different ways, using a variety of senses, and that certain forms of media are more effective than others [Gardner, 1992; Peart, 1984]. The preferred methods of delivery and learning include informal speeches by zookeepers, video display, interactive computers, and graphic exhibits [Gold & Benveniste, 1995]. Assessment of these various techniques is currently ongoing at Lincoln Park Zoo.

Modern technology has created the opportunity to include multi-sensory interactive exhibits, reflecting the different learning styles of the zoo visitor. The use of videotape is becoming a common means to demonstrate some of the more complex behaviors of the species exhibited. In addition, video is being used to show behind-the-scenes animal areas and management routines, as well as videotaped footage of primates in the wild. Some zoos are incorporating computer technology into their interpretive exhibits; touchscreen interactive computers provide large amounts of complex information to the motivated visitor. Computer games teach messages about primate behavior, natural history and conservation. Zoo Atlanta is experimenting with a virtual reality version of their gorilla exhibit, permitting the visitor to enter a 'virtual exhibit' and role play the part of an adolescent gorilla in the group (K. Burkes, personal communication).

Modern zoo signs have large colorful graphics and involve the visitor more by asking questions or directing the visitors' attention. Andersen [in press] wrote that the questions we ourselves ask are often the ones from which we learn the most. Through the design of functional and natural exhibits and the exhibition of appropriate social groupings, the animals will attract the visitors' attention and interest, generating many questions. If similar questions are then asked by nearby graphic displays, the visitor will be drawn to the graphics and learn more from them.

Over the past ten years more zoos have experimented with "interactive" signs that ask questions ("How much do they eat?") or give directions ("Compare the female and male lions."). By describing what the visitors are seeing, answering their questions, and asking or directing them to become active participants in their animal watching experiences, it is hoped that visitors will leave the exhibit more enriched than when they came. Zoos try to give visitors "an appetizer about animal life" and encourage them to take a closer look, thereby intensifying an interest in conservation [Andersen, 1995].

While a few zoological institutions (notably the New York Zoological Society, Chicago Zoological Society, and the Frankfurt Zoo) have long been involved in field research, many zoos are now beginning to fund field research or support active *in situ* conservation. Several examples are described in this volume. These partnerships with *in situ* research and conservation are often highlighted in the zoo's exhibits or interpretive displays, a very effective way to educate the zoo visitor about the zoo's commitment to conservation.

Other zoos have also incorporated ongoing scientific research into the design of their exhibitry, notably the Dallas Zoo's Gorilla Research Station, where visitors can view off-exhibit areas via closed-circuit television monitors that are used by zoo research staff. An innovative technique using a thematic approach to exhibit design is the Think Tank at the National Zoo in Washington, DC. It includes a laboratory in which orangutans display their cognitive capabilities using touch screen computers. Although the Think Tank highlights cutting edge research on apes' cognitive skills, it is somewhat similar to earlier public performances (described below), which demonstrated chimpanzee cognitive and motor coordination skills at the London Zoo [Morris, 1960].

Figure 6. A chimpanzee (*Pan troglodytes*) "tea party" at the Bronx Zoo, circa 1900. Although extremely popular entertainment, primate shows were phased out in the 1970s, as zoos developed an increased emphasis on conservation education. (Photographer unknown.)

PRIMATE SHOWS

In the early 1900s, several zoos began a tradition of training young apes (primarily chimpanzees, some orangutans and one notable gorilla) to perform publicly. Trained chimpanzee acts flourished from the 1920s through the 1970s, and a few wild animal parks in North America (at least one of which is an AZA-accredited institution) continue to this day. In the 1950s, several zoos (most notably the Detroit and St. Louis Zoos) staged elaborate performing chimpanzee shows, with apes riding horses, bicycles, motorbikes, and small cars, or performing acrobatic skills and playing music.

To illustrate the high intelligence of chimpanzees, the London Zoo began a series of daily public demonstrations of chimpanzee cognitive abilities. At set times in the morning and afternoon, the chimpanzees would "go through a series of tests, gymnastics, and intelligence problems for the dual benefit both of themselves and the visitors" [Morris, 1960, p. 21]. This included sorting through keys to open a locked cabinet for food, manipulating sticks in holes to acquire food, using ropes to pull food trays closer and swinging a tethered ball to knock grapes off pillars.

The extremely popular "tea parties" were staged for the amusement of zoo visitors (Figure 6) [see Wallis, this volume]. Tea parties continued at many institutions throughout the early and mid 1900s, lasting well into the 1970s. Morris [1960] questioned the message these chimpanzee shows provided the zoo visitor. He noted that for many years the chimpanzee tea party had been a popular feature of the summer season at the London Zoo, but its appeal had been anthropomorphic adding little to illuminate the true personality of the chimpanzee.

Cherfas [1984] recounted that chimpanzee tea parties finally ended at London Zoo in 1972 for many reasons. To be trained to perform, the chimpanzees were removed from their group and reared by human keepers. Upon reaching 6-8 years of age, they became unruly and were then reunited with their group (if possible). Many problems ensued due to human-rearing, and often the ex-performers did not fit in their former social group or the females often neglected their babies if they mated and delivered offspring. Despite the strong demand by the public to continue the tea parties, zoos began to concentrate on exhibiting social groups. Breeding and conservation became viewed as more important than entertainment. Although chimpanzee tea parties were enjoyed by both the public and the performers, in the long run they fostered a distorted image for the public and made social life and breeding difficult for the performers.

While most accredited zoos have ended the practice of primate performances, animal shows are still a mainstay at many zoos and aquaria, but primarily with marine mammals. Fortunately, these pinniped and cetacean shows are becoming increasingly conservation-minded in their scripting, educating visitors about animal adaptations, cognitive capabilities and natural behaviors. And, unlike in the primate shows of old, most of these animal performers can be kept in intact social groups.

CUTTING EDGE CONSERVATION EDUCATION AND FUTURE POSSIBILITIES

Some zoos are engaged in direct outreach, helping to fund reintroduction or research projects or aiding the establishment and maintenance of protected areas such as national parks. One unique approach is the Congo forest exhibit, under construction at the Wildlife Conservation Center, Bronx Zoo. A separate admission charge will be assessed to visit this series of exhibits and, upon exiting the complex, the visitor may choose from over 300 conservation projects funded by the Wildlife Conservation Society as the designated recipient of the small admission fee. Not only will this help to educate visitors about the various field research projects funded by WCS, but it is hoped that by providing visitors a choice and an opportunity to contribute, they will take pride in an active role in nature conservation.

Distance Learning – the use of video, satellite or internet computer technology to transmit information to and beyond the zoo site – is a promising area just beginning to be explored. Transmissions originating on zoo grounds are being sent into classrooms from institutions such as the Indianapolis Zoo and Zoo At-lanta. Future possibilities include live video feeds from remote *in situ* field sites (via satellite) to zoo interpretive centers, allowing bi-directional interactive learning possibilities. The zoo world is also developing an increased presence on the World Wide Web, with a strong slant toward conservation education [Appendix]. The use of the internet allows worldwide access to zoo animal information, exhibit interpretation and links with *in situ* conservation programs. The possibilities are rapidly expanding and still evolving as zoos are beginning to exploit this new media (see Lincoln Park Zoo's website at http://www.lpzoo.com, for an example of this technology).

CONCLUSIONS

No longer designed as a sterile display for wild animals, the modern zoo strives to establish a natural relationship between humans and our fellow crea-tures. Zoos want the public to know the origin of primates, what problems they face in the wild, and that their natural habitats have been shrinking and continue to do so. Through the design of naturalistic and functionally appropriate exhibits and enhanced interpretive methods, zoos contribute to the development of con-servation-aware citizens. It is the zoos' hope that visitors will become more knowl-edgeable about environmental matters and more willing to accept responsibility for their actions regarding the environment and, therefore, be motivated to help secure the future for nonhuman primates.

REFERENCES

Andersen, L.L. Exhibit design and interpretation in zoos. PROCEEDINGS OF THE EAZA ANNUAL GENERAL MEETING, Poznan, 1995 (in press).

Birney, B.A. Brookfield Zoo's 'Flying Walk' exhibit: Formative evaluation aids in the development of an interactive exhibit in an informal learning setting. ENVIRONMENT AND BEHAVIOR 20 (4): 416-434, 1988.

Blair, W. R. A Monkey Island and a New Bridge. BULLETIN, NEW YORK ZOO-LOGICAL SOCIETY Vol. XL #3, May-June, 1937.

Cherfas, J. ZOO 2000. London, British Broadcasting Corporation, 1984.

Coe, J.C. Design and perception: Making the zoo experience real. ZOO BIOL-OGY 4:197-208, 1985.

Coe, J.C.; Maple, T.L. Approaching Eden: A behavioral basis for great ape exhib-its. Pp. 117-128 in PROCEEDINGS OF THE AMERICAN ASSOCIATION OF ZOOLOGICAL PARKS AND AQUARIUMS ANNUAL CONFERENCE. Wheeling, West Virginia, American Association of Zoological Parks and Aquariums, 1984.

Dumond, F.V. The squirrel monkey in a seminatural environment. Pp. 87-146 in THE SQUIRREL MONKEY, L.A. Rosenblum; R.W. Cooper, eds. New York, Academic Press, 1968.

Evans, S.; Garber, P.; Green, S. The rainforest at Monkey Jungle: A natural laboratory. In RESEARCH IN ZOOS: PROSPECTS FOR PRIMATOLOGY. A.T.C. Feistner; S. Evans, eds. Jersey Wildlife Preservation Trust (in press).

Finley, T.; James, L.R.; Maple, T.L. People's perceptions of animals: the influence of the zoo environment. ENVIRONMENT AND BEHAVIOR 20(4):508-528, 1988.

Gardner, H. THE UNSCHOOLED MIND: HOW CHILDREN LEARN AND HOW SCHOOLS SHOULD TEACH. New York, Basic Books, 1992.

Goddard, D. SAVING WILDLIFE: A CENTURY OF CONSERVATION. New York, Wildlife Conservation Society, 1995.

Gold, K.C. Chimpanzee exhibits in zoological parks. Pp. 103-111 in CHIMPANZEE CONSERVATION AND PUBLIC HEALTH: ENVIRONMENTS FOR THE FUTURE. J. Erwin; J. Landon, eds. Rockville, Maryland, Diagnon/Bioqual, Inc., 1992.

Gold, K.C. Visitor attitudes at Lincoln Park Zoo. AMERICAN JOURNAL OF PRIMATOLOGY 36(2): 125, 1995.

Gold, K.C.; Benveniste, M. Visitors attitudes and behavior towards great apes. Pp. 152-158 in PROCEEDINGS OF THE AMERICAN ZOO AND AQUARIUM ASSOCIATION ANNUAL CONFERENCE. Wheeling, West Virginia, American Zoo and Aquarium Association, 1995.

Hancocks, D. ANIMALS AND ARCHITECTURE. London, Hugh Evelyn, 1971.

Hediger, H. Zoo architecture. Pp. 164-190 in BRITISH ZOOS: A STUDY OF ANIMALS IN CAPTIVITY. G. Schomberg, ed. London, Allan Wingate, 1957.

Hornaday, W.T. POPULAR OFFICIAL GUIDE TO THE NEW YORK ZOOLOGICAL PARK. New York, New York Zoological Society, 1913.

Koebner, L. ZOO BOOK: THE EVOLUTION OF WILDLIFE CONSERVATION CENTERS. New York, Forge Books, 1994.

Jones, G.R.; Coe, J.C.; Paulson, D.R. Woodland Park Zoo: Long Range Plan, Development Guidelines and Exhibit Scenarios. Jones and Jones Architectural Firm, Report for Seattle Department of Parks and Recreation, 1976.

Lang, E.M. The birth of a gorilla at Basle Zoo. INTERNATIONAL ZOO YEARBOOK #1, pp. 3-7, Zoological Society of London, 1960.

Loisel, G. The zoological gardens and establishments of Great Britain, Belgium, and the Netherlands. Pp. 407-448 in THE ANNUAL REPORT OF THE BOARD OF REGENTS OF THE SMITHSONIAN INSTITUTION, Washington, DC, Government Printing Office, 1908.

Mallinson, J.J.C.; Smith, D.; Darwent, M.; Carroll, J.B. The design of the Sumatran orangutan *Pongo pygmaeus abelli* 'home-habitat' at the Jersey Wildlife Preservation Trust. DODO 30:15-32, 1994.

Morris, D. The new chimpanzee den at the London Zoo. INTERNATIONAL ZOO YEARBOOK, #1 1959, Zoological Society of London, 1960.

Mottershead, G.S. Experiments with a chimpanzee colony at Chester Zoo. INTERNATIONAL ZOO YEARBOOK #1 1959, Zoological Society of London, 1960.

Normandia, S. Children's zoo design and evaluation. Pp. 475-480 in PROCEED-INGS OF THE AMERICAN ASSOCIATION OF ZOOLOGICAL PARKS AND AQUARIUMS ANNUAL CONFERENCE. Wheeling, West Virginia, American Association of Zoological Parks and Aquariums, 1986.

Ogden, J.J.; Lindburg, D.G.; Maple, T.M. Do you hear what I hear? The effect of auditory enrichment on zoo animals and visitors. Pp. 428-435 in PROCEED-INGS OF THE AMERICAN ASSOCIATION OF ZOOLOGICAL PARKS AND AQUARIUMS ANNUAL CONFERENCE. Wheeling, West Virginia, American Association of Zoological Parks and Aquariums, 1991.

Peart, B. Impact of exhibit type on knowledge gain, attitudes and behavior. CU-RATOR 17:220-237, 1984.

Process: Architecture V: 126. Jones and Jones: Ideas Migrate... Places Resonate. Tokyo, Japan, 1995.

Serrell, B. Visitor observation studies at museums, zoos and aquariums. Pp. 229-233 in PROCEEDINGS OF THE AMERICAN ASSOCIATION OF ZOOLOGI-CAL PARKS AND AQUARIUMS ANNUAL CONFERENCE. Wheeling, West Virginia, American Association of Zoological Parks and Aquariums, 1978.

Serrell, B. The evolution of educational graphics in zoos. ENVIRONMENT AND BEHAVIOR 20(4):396-415, 1988.

Seton, E.T. The National Zoo in Washington: a study of its animals in relation to their natural environment. Pp. 697-717 in THE ANNUAL REPORT OF THE SMITHSONIAN, Washington, DC, 1901.

Shettel-Neuber, J. Second and third generation zoo exhibits: A comparison of visitor, staff, and animal responses. ENVIRONMENT AND BEHAVIOR 20(4):452-473, 1988.

van den Bergh, W. The new ape house at Antwerp Zoo. INTERNATIONAL ZOO YEARBOOK, #1 1959, Zoological Society of London, 1960.

Wagner, R. AZA member statistical information. COMMUNIQUE August:23-24, 1996.

Dr. Kenneth Gold is the Director of Exhibit Interpretation at Lincoln Park Zoo in Chicago. He was formerly the Senior Research Associate for Zoo Atlanta's Conservation and Research Department. Dr. Gold has conducted many studies on captive primates throughout North America and has authored several papers on gorilla behavior and zoo exhibit design.

A black howling monkey (*Alouatta pigra*), endemic to Belize. (Photo by Robert Horwich.)

ZOOS AND *IN SITU* PRIMATE CONSERVATION

Fred W. Koontz

Bronx Zoo/Wildlife Conservation Society, Bronx, New York

INTRODUCTION

Species extinction may be taking place at the rate of 100 per day [Clark et al., 1994], an amount that is approximately 10,000 times greater than the natural background rate [Wilson, 1989]. Regardless of the exact rate, it is clear that humankind faces an environmental crisis. If this emergency is to be averted, many people and their businesses, cultural institutions, and governments must more directly participate in wildlife conservation. Because wild animal propagation skills and urban locations pre-adapt zoos as centers for nature preservation, it is not surprising that professionally-managed zoos were among the first institutions to respond to the environmental crisis by clearly focusing on wildlife conservation as their primary mission [Conway et al., 1984; Hutchins et al., 1991; Conway, 1995a].

For the last two decades, American zoos have aimed their conservation activities in four directions: animal breeding, public education, scientific research and fund raising [Conway, 1969, 1980; World Conservation Union, 1987; Kleiman, 1992; Wemmer & Thompson, 1995; Hardy, 1996]. In addition, zoo biologists have adopted ethics in agreement with modern conservation principles [Koontz, 1995]. Zoo budgets over the last 20 years have been restricted largely to operating expenses for animal propagation and for building animal habitat enclosures that are suitable for teaching ecology and inspiring the public to care about saving wildlife. Consequently, some biologists think that too little money has been directed by zoos toward *in situ* wildlife projects (e.g., habitat protection and other "on-the-ground," field-based, actions) [e.g., Eudey, 1995]. Until recently, zoo biologists generally responded to this criticism by citing the importance of *ex situ* wildlife conservation (e.g., environmental education and conservation breeding programs for reintroduction), and by explaining how zoo-based activities indirectly assist wild animal survival in nature [e.g., Seal, 1988]. In the last several years, however, zoos have increased significantly their *in situ* wildlife conservation efforts [Conway, 1989; Hutchins & Wiese, 1991; Conway, 1995a, 1995b; Hutchins & Conway, 1995].

The American Zoo and Aquarium Association (AZA), headquartered in Bethesda, Maryland, is the professional organization of American zoo and aquarium personnel. In 1981, AZA members proclaimed that wildlife conservation was their primary mission. Since that time, AZA conservation efforts have grown steadily. In the last year, for example, over 450 wildlife conservation projects in more than 65 countries were undertaken by the 162 accredited AZA institutions, and its members produced more than 400 technical publications [American Zoo and Aquarium Association, in press]. This chapter discusses the *in situ* conservation role of AZA zoos today, and their likely role in the future. It is hoped that a better understanding of zoo activities by primatologists will strengthen the potential for future partnerships, and ultimately improve cooperation that might lead to an increased number of primate conservation projects.

THE EVOLUTION OF ZOOS TOWARD WILDLIFE CONSERVATION PARKS

Although people have been keeping collections of wild animals since at least 2300 BC, it was not until the mid-1800s that zoos began to appear in Europe [Koebner, 1994]. Over the course of the last 100 years, the role of zoos within human society has followed an evolutionary course that, in many ways, reflects the attitudes, values, and curiosities that people of different times have displayed toward wild animals [Rabb, cited by Koebner, 1994; and see Wallis, this volume]. From the turn of the 20th Century until the early 1960s, American and European zoos were largely "wild animal menageries." The theme was taxonomic. The idea was to display animals in a manner to allow easy observation and comparison, with a special emphasis on displaying rare and unusual animals so the wide variety of wildlife found within the animal kingdom could be appreciated.

In the 1960s and 1970s, as the concepts associated with ecology took hold in the public's mind, American zoos evolved away from menageries into "living natural history museums." A shift toward establishing formal educational programs, and building modern, habitat-style enclosures (which actually started many years before in Europe) that display animals in more natural settings became the objective of most zoo directors [see Gold, this volume]. It was during this period that zoo exhibitry became much more sophisticated, partially as a response to new building materials (e.g., epoxy resins and Fiberglas) and construction methods [Sausman, 1982]. At this same time, it became apparent to leading zoo curators that sources of animals in the wild were decreasing, and as a result, cooperative breeding of animals was not only advantageous from a conservation point of view but necessary to maintain zoo stocks [Conway, 1974; Ralls et al., 1979]. The idea of zoos serving their communities not only as recreational facilities and educational museums, but also as conservation centers, had its origin during this period.

In the 1980s and 1990s, American zoos evolved again, this time in a direction that they continue to move: toward becoming "environmental resource centers" or "wildlife conservation parks." Education programs at zoos began to focus on environmental health as their primary message. Organized, cooperative, long-term (50-200 years) animal breeding programs, called Species Survival Plans© (SSPs), were initiated by the AZA in the early 1980s. These programs grew steadily and were supplemented in the mid- and late-1980s with a series of AZA support committees: the Taxon Advisory Groups (TAGs), Scientific Advisory Groups (SAGs) and Faunal Interest Groups (FIGs). Since 1993, the AZA has had a Field Conservation Committee to assist zoos with the implementation of *in situ* conservation projects. These committees are composed of zoo professionals and technical advisors from other organizations.

By 1996, 162 zoos and aquariums were accredited by the AZA and participating in its wildlife conservation programs. Also, in the last 15 years American zoo biologists reached out to other world regions in an effort to develop global, zoo-based conservation programs; this work was accomplished primarily by the Conservation Breeding Specialist Group (CBSG) of the World Conservation Union (IUCN).

The idea of evolving zoos more strongly as conservation resource centers or "wildlife conservation parks" is only now coming to fruition [Conway, 1995a, 1995b]. These new institutions not only possess the traditional facilities for keeping and displaying wild animals (and the long-term propagation and educational programs that go along with them), but as environmental centers they also directly support *in situ* conservation projects by providing administrative, curatorial, educational, financial, scientific and veterinary resources [Hutchins & Conway, 1995]. This support is made, for example, by sending zoo-based personnel to work in the field, sharing animal and scientific facilities, donating money for *in situ* work and contributing animals for reintroduction. In addition, linking zoo animal collections more directly with *in situ* programs is becoming more commonplace [Conway, 1995a]. For example, the Bronx Zoo/Wildlife Conservation Society elephants were used as models for developing the satellite technology necessary for field biologists to track forest elephants in Cameroon [Koontz, 1992].

Despite the promising trend that zoos are contributing toward wildlife preservation, including *in situ* conservation, zoo critics still question how significant this conservation role can be, and should be, in the future [e.g., see Norton et al., 1995; Snyder et al., 1996]. Zoo limitations have been noted based on ethical, financial, political, scientific and technical grounds. For example, Snyder et al. [1996] pointed out some serious problems with captive breeding of endangered species (e.g., high costs, disease outbreaks, domestication, lack of administrative continuity, and poor success of reintroduction programs to date) [but see Hutchins et al., 1997]. It seems likely that the debate will continue, especially regarding the *in situ* conservation responsibility of zoos.

ZOO PERSONNEL AND *IN SITU* CONSERVATION

Operating a modern zoo requires a diverse team of professionals and sophisticated animal care facilities, including scientific laboratories [Sausman, 1982; Kleiman et al., 1996]. Experts in animal behavior, conservation genetics, education, endocrinology, exhibit design, population biology, nutrition, management, reproductive physiology, systematics, pathology and veterinary medicine, among others, work together with specialists concerned with caring for amphibians, birds, fishes, invertebrates, mammals and reptiles. Besides serving as urban wildlife education centers, the human and material resources of zoos make them ideally suited for providing support services for field conservation projects, as the skills and equipment required to care for *ex situ* and *in situ* wildlife overlap.

Zoos employ a variety of experts that are considered here as "zoo biologists." Many *in situ* conservation projects are undertaken each year by specialists working for zoos, usually as curators, who study amphibians, birds, fish, mammals and reptiles [e.g., see AZA Annual and Regional Conference Reports and the *International Zoo Yearbook*]. However, in addition to curators, zoos staff include conservation geneticists, endocrinologists, ethologists, nutritionists, population biologists, systematicists and reproductive physiologists. These zoo biologists have the potential to work collaboratively with field primatologists to solve complex conservation problems. For example, zoo population biologists have developed much of the theory and software needed for genetic and demographic management of small populations [e.g., Ballou et al., 1995]; their work began on zoo populations but is now relevant for managing small, fragmented wild populations [e.g., Lacy, in press]. The National Zoological Park (Washington, DC), Brookfield Zoo (Chicago, Illinois), Lincoln Park Zoo (Chicago, Illinois), Toledo (Ohio) Zoo, AZA's Conservation Center (Bethesda, Maryland) and the Conservation Breeding Specialist Group (Minneapolis, Minnesota) are all cited by zoo biologists for their significant work in population biology [Ballou et al., 1995].

Reproductive physiologists and endocrinologists working at zoos have contributed greatly to techniques of cryobiology, embryo transfer, hormonal analyses and artificial insemination [Dresser, 1986; Czekala et al., 1986; Asa and Sarri, 1991; Asa, 1996]. All of these techniques may be needed for management of small, isolated wild populations in the near future [Conway, 1995b]. The National Zoological Park, Cincinnati Zoo, San Diego Zoo, St. Louis Zoo, Bronx Zoo/Wildlife Conservation Society, and Woodland Park Zoo (Seattle, Washington) are known for their efforts in these areas.

Conservation genetics is an area of special concern for zoos, because of the importance of fully managing the genetics of zoo populations and clearly understanding the systematics of animals in zoo collections. The San Diego Zoo, Bronx Zoo/Wildlife Conservation Society and National Zoological Park possess molecular genetics programs that have contributed significantly to conservation biology by providing both practical advise and theory to field biologists [e.g., Ryder, 1986; Amato & Gatesy, 1994].

Nutrition is now recognized in the zoo profession as a critical factor in successfully maintaining wild animals in captivity [Allen, 1996]. Yet, it has only been in the last 20 years that formal nutrition laboratories and research programs have been launched by American zoos. In North America, the first zoos to hire professional nutritionists were Metro Toronto Zoo in 1975, the National Zoological Park in 1978, Brookfield Zoo in 1980, and the Bronx Zoo/Wildlife Conservation Society in 1985. Despite this relatively recent start, zoo nutritionists have worked with primatologists to contribute important information to the scientific literature [e.g., Dierenfeld et al., 1992; Oftedal & Allen, 1996]. These laboratories, most equipped with the ability to quantify levels of proteins, fats, carbohydrates, minerals and vitamins, are now undertaking collaborative research with field biologists to investigate feeding ecology of free-ranging animals [e.g., Yeager et al., 1997].

Zoo veterinarians, including pathologists and reproductive specialists, have a wide base of experience and technical expertise gained from working with a great diversity of zoo animals. Their hands-on skills and knowledge can be applied usefully to many field situations [Hutchins et al., 1991]. Zoo veterinarians have the ability to test methodologies before applying them in nature; for example, much of the development work for learning how to chemically immobilize wild animals took place in zoos [e.g. Bush, 1992, 1996]. Since 1989, the Bronx Zoo/Wildlife Conservation Society has maintained a Field Veterinary Program, with a full-time wildlife veterinarian assigned to assist field biologists with *in situ* projects [Karesh & Cook, 1995, in press]. Many other zoos have also sent their zoo veterinarians into the field; special emphasis has been placed on providing help with chemical immobilizations, disease surveys and tissue collection. Not only have clinical veterinarians taken part in these efforts, but AZA specialists in pathology and reproductive assistance have also worked on many *in situ* projects in recent years. AZA veterinarians have also provided help to stranded marine mammals and many other cases of injured wildlife.

Because zoos have large collections of live animals in captivity, including many endangered and threatened species, they provide an opportunity to conduct research with these animals that would be difficult – sometimes impossible – under field conditions [for a review of zoo research programs, see Hardy, 1996]. The resulting knowledge gained can be directed toward conservation of these and similar species in nature.

Most zoos have active wildlife education programs and the necessary infrastructure to develop and produce educational materials, as supplements to *in situ* conservation projects. For example, for many years the Fort Wayne (Indiana) Children's Zoo provided funding and technical assistance for Richard Tenaza's education efforts in Indonesia for the Mentawai Island primates. Zoos also export their educators to help with environmental education around the globe [e.g., Savage et al., this volume]. The Bronx Zoo/Wildlife Conservation Society, for example, has sent several of its educators to Belize, China, and Venezuela to instruct teachers on methods of wildlife education. The potential is great for zoo educators to expand such work with *in situ* wildlife park educators in the near future.

Zoo biologists have also provided professional training for wildlife biologists working in less developed countries. The National Zoological Park's Conservation and Research Center (Front Royal, Virginia) has for many years led the way in this effort [e.g., Wemmer et al., 1993]. Additional zoos, such as the Bronx Zoo/ Wildlife Conservation Society, Brookfield Zoo, Milwaukee County Zoological Park, White Oak Plantation (Yulee, Florida), and Fossil Rim Wildlife Center (Texas) also have participated in international training programs.

It is not only zoo biologists, veterinarians and zoo educators that might offer assistance to field conservation: zoo administrators, business managers, exhibit designers and public relations personnel can help international wildlife park staffs with many activities. While there are few examples to date, it appears that such efforts will soon become more routine. Several zoos have developed "adopt-a-park" programs to foster such relationships. Most notable of these is the program of the Minnesota Zoo working in Java, Indonesia. In many parts of the world, wildlife parks operate on small budgets and with minimal support and professional expertise. In the future, American zoos may collaborate with wildlife parks to offer administrative and business assistance, as well as wildlife management consultation.

COMMON FEATURES OF WILDLIFE MANAGEMENT
AND ZOO ANIMAL MANAGEMENT

Wildlife managers require many of the same resource needs as zoo managers. These include: animal handling skills, animal nutrition knowledge, behavior observation skills, facilities to produce graphic art and signs, genetics and systematics analyses, population biology modeling, professional training opportunities, rapid pathology diagnostics, reproductive enhancement and contraception, wildlife education expertise and veterinary services [Conway, 1989, 1995b].

Accelerating habitat destruction and fragmentation suggests that the differences between zoo and wild habitats will continue to narrow in the future; many wildlife reserves will become increasingly "zoo-like," e.g., they will have perimeter fences and contain small populations of endangered animals [Conway, 1989]. It is likely, therefore, that wildlife management in nature will require ever greater amounts of human intervention and manipulation. Zoo personnel and their associated resources are prepared to assist in such efforts. To fully participate in wildlife conservation efforts, it is incumbent upon zoo personnel to work in partnership with field-based conservationists and to share resources as much as possible.

Zoo-based primatologists conduct important conservation projects in nature with wildlife managers. Their skills and expertise tend to be multi-disciplinary in scope: animal handling, behavioral analyses, conservation genetics, feeding ecology, population biology, restoration biology and veterinary care are typical applications. An example of a recent *in situ* primate conservation project is described in the following section. The project, the translocation of black howler monkeys in Belize, incorporated all of these zoo-related skills and illustrates how zoo personnel can work with field primatologists.

Relatively few primate translocations have been systematically studied, and as a result, the effectiveness of translocation as a conservation method for re-establishing primate populations is largely untested [for review, see Caldecott & Kavanagh, 1983; de Vries, 1991]. Unfortunately, the declining status of most primates, coupled with an increasingly fragmented world, suggests that future wildlife managers will be required to translocate primates and other animals. Through their experiences handling and managing animals in zoos, zoo-based biologists and veterinarians are well prepared to conduct these translocations. It is important, therefore, that they begin now to devise protocols and evaluate methods by undertaking translocation projects, when appropriate.

CASE STUDY: REINTRODUCTION OF BLACK HOWLING MONKEYS IN BELIZE

A reintroduction of black howling monkeys *(Alouatta pigra)* (Figure 1) into the Cockscomb Basin Wildlife Sanctuary (located near Maya Center, Belize) was completed by a team of conservationists that brought together field primatologists, wildlife park managers and zoo biologists. The project included capturing, moving, temporarily holding and then releasing 62 free-ranging howlers between May, 1992, and May, 1994. Thereafter, the translocated monkeys were studied continuously until June, 1996 [for additional details, see Horwich et al., 1993; Koontz, 1993, in press; Koontz et al., 1994].

This species restoration effort was undertaken because black howlers, which historically lived in the Cockscomb Basin, became locally extinct in 1978 as a cumulative effect of a yellow fever epidemic (1957-1959), a hurricane (1961) and over-hunting (1963-1978). By 1991, hunting was controlled and yellow fever had disappeared, but there was little chance of natural re-colonization because of few howlers in the area and the fact that these were blocked from entering the Cockscomb Basin because of intervening mountains. After consulting the IUCN's Re-introduction Criteria, it was decided that translocation in this case was the best alternative for species restoration.

The project was carried out by a team led by field primatologist Robert Horwich (Community Conservation Consultants), Cockscomb Basin Wildlife Sanctuary Manager Ernesto Saqui (Belize Audubon Society), and the author. Co-operation among many additional individuals and institutions was essential for this project. Major collaborators included: primate capture expert, Ken Glander (Duke University Primate Center); wildlife veterinarian, Wendy Westrom (Bear Mountain Veterinary Service); and park wardens, Hermelindo Saqui and Emiliano Pop. Other project participants included: the Bronx Zoo's General Curator James Doherty, Mammal Curator Patrick Thomas and Collections Manager Penny Kalk; Belize Audubon Society's Protected Areas Director, Osmany Salas; and computer consultant, Charles Koontz (Columbia Consultants). Fordham University and Wild-

Figure 1. Black howling monkeys (*Alouatta pigra*) from the community Baboon Sanctuary were translocated to the Cockscomb Basin Wildlife Sanctuary (CBWS) as part of a species restoration plan. (Photo by Ken Gold.)

life Conservation Society Science Resource Center predoctoral students Linde Ostro and Scott Silver joined the research effort in 1994, after working for several years as Bronx Zoo animal keepers.

As there were no howlers on public lands in Belize in 1991, translocation of monkeys to Cockscomb Basin Wildlife Sanctuary (CBWS), one of Belize's premier wildlife parks, was considered conservationally important in order to establish a new, protected population. Howlers are ideal candidates for translocation; they are folivorous, live in small groups, and occupy small home ranges [Horwich & Lyon, 1990]. The project's primary goal was to re-establish a population of howlers in the CBWS by introducing founders from the Community Baboon Sanctuary, a private reserve located 100 km north near Bermudian Landing [Horwich & Lyon, 1990]. The release site's howler habitat (>20,000 ha) was estimated to be able to support at least 5,000 howlers. The proposal called for translocating 60-65 monkeys (12-15 troops) over two years. Objectives included: (1) studying and improving primate translocation, especially the monkeys' acclimation, survival and reproduction; (2) increasing ecotourism in CBWS by placing primates near visitor trails; (3) training CBWS managers in scientific study; (4) providing additional financial support for the Belize Audubon Society; and (5) increasing public appreciation of wildlife in Belize, especially of black howlers.

In March, 1991, the Project Directors surveyed the vicinity of Cockscomb Basin. They found few howlers and a great amount of habitat conversion. Discussions with Belizean officials indicated their support for the project and all necessary permits were obtained. Evaluation of the proposed release site revealed that it was appropriate, because: (1) it was within the historic range; (2) was relatively large and protected; (3) no other primate species was living there (minimizing disease transmission risk); (4) the park was surrounded by mountains (minimizing dispersal risk); (5) wardens were available for monitoring the released monkeys; (6) vehicles could get to within several km of the release sites; and (7) park headquarters there included lodging, radio communication facilities, and a trail system for ready access into the forest. Pre-project activities included establishing a 5 km by 5 km study area that was centered about the park headquarters and mapping 20 km of trails for post-release monitoring. Because fruit is most abundant and the roads are most passable during the late dry season (January - May), the team determined that the translocations should take place during May.

In May, 1992, the team conducted a pilot study by capturing, moving, holding for three days, and releasing 14 black howlers, comprising three social groups (see below for a discussion of method details). All adult translocated howlers were followed by radio tracking (females wore necklace transmitters and males carried ankle bracelet transmitters or implanted radios) for one year. All fourteen of these pilot study animals survived 30 days, and 12 (86%) survived at least one year. Dispersal from release sites was minimal, feeding behavior seemed normal and reproduction occurred. Only after this pilot study showed such promising results did the team decide to proceed with the translocation plan, in which 48 additional animals were moved, in May, 1993, and May, 1994. Thus, a total of 62 animals from 14 troops were introduced into the Cockscomb Basin Wildlife Sanctuary.

Troops for translocation were chosen carefully for appropriate demographic traits (e.g., males and females present; no animals between six months and 18 months of age, as they were too difficult to dart). Howlers were captured by darting each monkey with the anesthetic Telazol. After a monkey was drugged, it usually fell into a net within several minutes. Entire troops were usually caught in this manner, one monkey at a time (12 of 14 groups were complete and two groups were mixed-sexed subsets of larger troops). In 1994, four troops were studied for three months prior to capture by Scott Silver and Linde Ostro, so that pre- and post-release behavior could be compared.

At capture, all monkeys were: (1) given physical exams; (2) measured, weighed and aged by tooth wear patterns; (3) marked for permanent identification with transponder chips and colored ankle bracelets; (4) fitted with radio telemetry (adults only); (5) treated for minor wounds; (6) bled for later hematology, serum chemistries, nutritional variables (e.g., minerals and vitamins), genetic variation, and to establish a blood bank for disease study, if needed; and (7) checked for fecal parasites. One infant was rejected by its mother during the capture process and is now living at the Belize Zoo.

Monkeys were placed in individual crates and transported to Cockscomb Basin Wildlife Sanctuary by either helicopter (45 minutes) or by air-conditioned van (four hours). Monkeys in four troops were hard released in the late afternoon by simply taking them into the forest and letting them go. Ten troops for soft release were carried into the forest immediately after arriving in Cockscomb and placed in holding cages. The soft release cages measured 4.92 m x 2.46 m x 3.08 m (length x width x height), and were made of lumber and 2.5 cm chicken wire. These cages were constructed several months in advance of the translocation. Sites for soft release cages (and for hard release locations) were chosen to be about 1 km from other howlers and at the base of a food tree. The soft release groups were held for 24 to 72 hours, during which time they were provided water, available fruit and fresh browse. All releases took place between 1500 and 1600 hours, based on our assumption that an afternoon release might help prevent panic dispersal; as night fell, the howlers would likely stop traveling to sleep. The monkeys were let out of their cages by cutting a large hole in the wire screen. In all cases, the animals calmly exited and we believe stayed within 100 meters of their release site for at least 24 hours.

In summary, between May, 1992, and May, 1994, 62 howlers (28 males, 34 females), of which 51 were adults (22 males, 29 females), were released into the CBWS. As of June, 1996, minimum survival statistics for all translocated howlers are estimated as: (1) 95% survived capture and were released; (2) 93% survived 30 days; (3) 84% survived at least six months; and (4) 80% survived one year. The median movement from release sites was estimated at less than 2 km, but 15 animals (24%) moved 5-9 km. Thus, 60% of the translocated howlers both survived long-term and stayed near the release sites. Home ranges seemed to be established during the first six months, however, two troops made large changes in location between six and eight months after release. Home range core areas were much larger (average 57 ha) than those reported for howlers living at the Community Baboon Sanctuary (1-10 ha). Twenty-four per cent of the released monkeys (including both juvenile and adults) deserted their original troops, but eventually entered other troops; similar dispersal has been noted by Glander [1992] for non-translocated mantled howlers (*A. palliata*). Over 30 infants were born in the first four years of the project, with 70% surviving. No significant differences in translocation responses were detected between those howlers that were hard or soft released, however, the project team recommends the soft release methods as a more conservative approach.

The translocated howlers have shown high survivability (at least 80% after one year), low dispersal (0-9 km, mostly < 3 km), and high infant survivability (70%). The most promising statistic, however, is that the core population of 37 translocated howlers (nine troops) that established residence near the visitor trails (and are the most accurately monitored), have multiplied to at least 60 (about 15% population growth per year). Also, wildlife conservation awareness was improved in Belize by: (1) inviting Belizean television and radio crews to participate; (2) holding public lectures; (3) publishing yearly articles in the Belize Audubon Society's Newsletter; and (4) setting up a display board in the CBWS's visitor center.

The four-year budget (1992-1996) for the project was $99,000.00. This translates into a "cost per reintroduced monkey" of $1597.00. This figure includes all team expenses for travel and lodging, all supplies and equipment, and post-release monitoring to date. The budget included salaries for most of the local staff, but the salaries of the USA-based team members were donated by their respective institutions. Helicopter time was donated by the British Royal Air Force.

The preliminary results of this on-going study with black howlers is promising. The CBWS black howler population estimate for January, 1997, is 80-100 monkeys. While it is too soon to make any final conclusions, the project team believes that, with some demographic and stochastic luck, the black howlers now living in the Cockscomb Basin have a reasonable chance to re-establish a viable population. We found that many of the skills needed for primate translocation are routine practices for persons working with zoo primates.

AZA PRIMATOLOGISTS WORKING *IN SITU*

The howler case study is just one example of AZA biologists working *in situ*; there are many similar cases each year. The more publicized efforts usually involve reintroductions and translocations [for review, see Jones, 1990; Gipps, 1991; Kleiman, 1996]. The most significant primate reintroduction to date by zoo biologists has been the work of Kleiman, Beck, and their associates [e.g., Kleiman et al., 1986, 1991; Stoinski et al., this volume], who have for over a decade conducted much of the ground-breaking work in Brazil for the reintroduction and "metapopulation management" (discrete populations, including the captive population, being managed as one unit) of golden lion tamarins (*Leontopithecus rosalia rosalia*).

Anne Savage, the cotton-top tamarin AZA SSP Coordinator based at Roger Williams Park Zoo (Providence, Rhode Island), has contributed significantly to field primatology by her excellent studies of the cotton-top tamarin (*Saguinus oedipus oedipus*) [e.g., Savage et al., 1993, 1996, this volume]. Examples of other zoo-based primatologists who have worked in the field include: Jeanne Altmann (Brookfield Zoo), Don Lindburg (San Diego Zoo); Ron Tilson (Minnesota Zoo); Anne Baker (Syacuse Zoo); Andy Baker (Philadelphia Zoo) and Colleen McCann (Bronx Zoo). In addition, many members of the AZA's Primate Taxon Advisory Groups are now planning *in situ* conservation projects. Chairpersons of these TAGs are: Ingrid Porton (Prosimians, St. Louis Zoo); Anne Baker (New World Monkeys, Syracuse Zoo); Andy Baker (New World Monkeys, Philadelphia Zoo); Fred Koontz (Old World Monkeys, Bronx Zoo); Eve Watts (Old World Monkeys, San Francisco Zoo); Dave Ruhter (Old World Monkeys, Silver Springs, Florida); Ron Tilson, (Gibbons, Minnesota Zoo); and Terry Maple (Great Apes, Zoo Atlanta). Many of these individuals are contributors to this volume.

FUND RAISING AT ZOOS FOR *IN SITU* CONSERVATION

Over 100 million people visit North American zoos and aquariums each year. Such large numbers of people offer hope that new and significant amounts of money can be raised for *in situ* wildlife conservation projects. This potential to date, however, has been largely unrealized. For example, by simply implementing a $1.00 conservation surcharge, AZA zoos could provide a much needed reliable source of conservation support, but much time will have to be devoted to the education of local authorities to win approval for such a program [Conway, 1995a]. This idea is hampered by the fact that many zoos are municipal, city parks, and as such are not privately owned or managed by non-profit organizations; they may be restricted in their fund raising efforts by local legislation and zoo rules. However, the trend is for city-run zoos to be turned over to non-profit organizations, who have more flexibility in financial arrangements.

I believe that eventually zoos and aquariums will spend 5% to 15% of their operating budgets on *in situ* conservation. Nevertheless, even today, zoos and aquariums are contributing significant sums of money and personnel time to field projects. For example, the Bronx Zoo/Wildlife Conservation Society spent nearly $12 million (15% of its annual budget) last year on field conservation efforts. It has been suggested by some zoo professionals, but not approved to date, that a zoo applying to become an accredited "AZA Wildlife Conservation Park" will be required to spend a minimum amount directly on field conservation. I predict that, eventually, this requirement will be adopted.

ZOO ANIMALS AND REINTRODUCTION

In the 1980s, zoos oriented much of their long-term propagation programs toward the idea of eventual reintroduction of animals to nature. In some cases it was planned that the reintroduction might not occur for 100 or more years. For many zoos, these long-term programs became their primary *in situ* conservation focus. While these efforts may be important in future restoration projects, this effort hardly counts as current *in situ* conservation activity. This kind of thinking has been described by zoo biologists as the "Noah's Ark Paradigm." However, in the early 1990s a shift in attitudes began to take hold in AZA zoos; while it was agreed that reintroduction of zoo-born animals into the wild may be an important tool [see Snyder et al., 1996], it became increasingly evident to many zoo biologists that it is only one small piece of what needs to be done. In the midst of a species extinction crisis, zoos can not afford to wait 100 years to participate in field efforts. What is needed is for many more zoos to develop a more holistic approach to zoo-based conservation. Thus, the idea of the "Wildlife Conservation Park" or zoos becoming "environmental resource centers" has emerged [Conway, 1995a].

Re-invigorated and expanded education programs, science research programs, fund raising efforts, and field projects have been the trend for the last five years in AZA zoos, as they have moved to become more holistic. What remains to be seen, however, is if this trend will continue. For zoos to reach their full potential, it will be necessary for them to form partnerships with other institutions and scientists (e.g., natural history museums, universities, businesses, and individuals). For example, at the Bronx Zoo/Wildlife Conservation Society, we recently joined a "conservation and research consortium" with Columbia University, the American Museum of Natural History, the New York Botanical Garden and the Wildlife Preservation Trust International. By combining resources, new programs, projects and funding opportunities have become possible. The consortium hopes that ultimately the winner will be wildlife.

LINKING ZOO PROGRAMS WITH *IN SITU* CONSERVATION

Since the early 1980s, the AZA and the CBSG have improved organizational strategies and procedures (mostly by establishing various committees and special workshops) that have helped zoo biologists to better coordinate animal collection planning and other cooperative programs; the intent always has been to maximize the contribution that zoos can make to wildlife conservation [e.g., Stevenson et al., 1992; Hutchins et al., 1995]. A clear objective of these efforts has been to link the needs of wild animals with zoo programs, and to encourage cooperation between zoo personnel and field conservation biologists.

The AZA, through its Taxon Advisory Groups, is now developing Regional Collection Plans ("regional" referring to North American zoos). These plans not only list the suggested taxa to be held in AZA zoos, but suggest possible *in situ* projects for zoos to consider [e.g., Koontz, 1994; Baker, 1994; Thompson-Handler et al., 1995]. In addition, the AZA's Faunal Interest Groups help to identify *in situ* projects for zoo participation. Both the AZA and CBSG routinely ask field scientists to review resulting documents, in hope that the work's relevance and importance to field conditions will be improved.

I suggest here that greater communication and collaboration between zoo biologists and other primatologists would be beneficial to AZA programs and primate conservation. How can this be accomplished? More regular attendance at primate meetings by zoo personnel is needed. Primatologists need to make themselves aware of the changing direction in AZA zoos today, especially regarding animal collection planning and field conservation activities. Also, it is important for primatologists to understand that zoos have considerable challenges to face, and as a result, they can not evolve as quickly as some would hope. Nevertheless, zoo contributions to *in situ* conservation are increasing each year and can no longer be ignored – especially if the work of AZA zoos is considered collectively.

CONCLUSIONS

Zoos evolve to meet the changing needs of the communities they serve. As we enter the next century, humankind must place greater effort on wildlife conservation problems. Zoos are prepared to become new kinds of institutions that provide financial, human and material resources to support national and international wildlife conservation education programs, research projects, and *in situ* conservation activities. In fact, AZA zoos are already evolving in this direction; this new kind of zoo is called a "Wildlife Conservation Park" [Conway, 1995a]. While it is true that few zoos today have significant *in situ* programs, when considered collectively, the 162 AZA zoos are making a significant impact in the field. In addition, the trend for growth in this area by American zoos is steeply positive [Hutchins & Conway, 1995]. Finally, while this chapter has a North American and AZA bias, other zoo associations around the world are now forming similar cooperative wildlife conservation programs, both *ex situ* and *in situ*.

I suggest that primatologists from academia and other organizations other than zoos – especially those working on field conservation projects – consider collaborating on AZA or similar zoo-association primate programs. It is most important that perceived and real past differences in conservation approaches must not prevent zoo and field primatologists from working together in the future; working together, we can produce more holistic conservation strategies for endangered and threatened primates.

REFERENCES

Allen, M.E. Nutrition: Introduction. Pp. 107-108 in WILD MAMMALS IN CAPTIVITY: PRINCIPLES AND TECHNIQUES. D.G. Kleiman; M.E. Allen; K.V. Thompson; S. Lumpkin, eds. Chicago, University of Chicago Press, 1996.

American Zoo and Aquarium Association (AZA). 1996 ANNUAL REPORT ON CONSERVATION AND SCIENCE. Bethesda, Maryland, American Zoo and Aquarium Association, (in press).

Amato, G.; Gatesy, J. PCR assays of variable nucleotide sites for identification of conservation units. Pp. 215-226 in MOLECULAR ECOLOGY AND EVOLUTION: APPROACHES AND APPLICATIONS. B. Schierwater; B. Streit; G.P. Wagner; R. DeSalle, eds. Basel, Switzerland, Birkhauser, 1994.

Asa, C. Reproductive physiology. Pp. 390-417 in WILD MAMMALS IN CAPTIVITY: PRINCIPLES AND TECHNIQUES. D.G. Kleiman; M.E. Allen; K.V. Thompson; S. Lumpkin, eds. Chicago, University of Chicago Press, 1996.

Asa, C.S.; Sarri, K.J. Assessment of biotelemetry for monitoring reproductive events related to changes in body temperature and activity rhythm in cheetahs and spider monkeys. Pp. 49-57 in BIOTELEMETRY APPLICATIONS TO CAPTIVE ANIMAL CARE AND RESEARCH. C.S. Asa, ed. Wheeling, West Virginia, American Association of Zoological Parks and Aquariums, 1991.

Baker, A., ed. AZA NEW WORLD MONKEY REGIONAL COLLECTION PLAN. Syracuse, New York, Burnet Park Zoo, 1994.

Ballou, J.D.; Gilpin, M.; Foose, T.J., eds. POPULATION MANAGEMENT FOR SURVIVAL AND RECOVERY: ANALYTICAL METHODS AND STRATEGIES IN SMALL POPULATION CONSERVATION. New York, Columbia University Press, 1995.

Bush, M. Remote drug delivery systems: Review article. JOURNAL OF ZOO AND WILDLIFE MEDICINE 23:159-180, 1992.

Bush, M. Methods of capture, handling and anesthesia. Pp. 25-40 in WILD MAMMALS IN CAPTIVITY: PRINCIPLES AND TECHNIQUES. D.G. Kleiman; M.E. Allen; K.V. Thompson; S. Lumpkin, eds. Chicago, University of Chicago Press, 1996.

Caldecott, J.; Kavanagh, M. Can translocation help wild primates? ORYX 17:135-139, 1983.

Clark, T.W.; Reading, R.P.; Clark, A.L. ENDANGERED SPECIES RECOVERY: FINDING LESSONS, IMPROVING THE PROCESS. Washington, DC, Island Press, 1994.

Conway, W.G. Zoos: their changing roles. SCIENCE 161:48-52, 1969.

Conway, W.G. Animal management models and long-term captive propagation. Pp. 141-148 in PROCEEDINGS OF THE AMERICAN ASSOCIATION OF ZOOLOGICAL PARKS AND AQUARIUMS ANNUAL CONFERENCE. Wheeling, West Virginia, American Association of Zolligcal Parks and Aquariums, 1974.

Conway, W.G. An overview of captive propagation. Pp. 199-208 in CONSERVATION BIOLOGY: AN EVOLUTIONARY-ECOLOGICAL PERSPECTIVE. M.E. Soule; B.A. Wilcox, eds. Sunderland, Massachusetts, Sinauer Associates, 1980.

Conway, W.G. Miniparks and megazoos: From protecting ecosystems to saving species. THE THOMAS HALL LECTURE, Washington University, St. Louis, Missouri, February 28, 1989.

Conway, W.G. The conservation park: a new synthesis for a changed world. Pp. 259-276 in THE ARK EVOLVING: ZOOS AND AQUARIUMS IN TRANSITION. C. Wemmer, ed. Front Royal, Virginia, Conservation and Research Center (Smithsonian Institution), 1995a.

Conway, W. Wild and zoo animal interactive management and habitat conservation. BIODIVERSITY AND CONSERVATION 4:573-594, 1995b.

Conway, W.; Foose, T.; Wagner, R. SPECIES SURVIVAL PLAN OF THE AMERICAN ASSOCIATION OF ZOOLOGICAL PARKS AND AQUARIUMS. Wheeling, West Virginia, American Association of Zoological Parks and Aquariums, 1984.

Czekala, N.M.; Gallusser, S.; Meier, J.E.; Lasley, B.L. The development and application of an enzyme immunoassay for urinary estrone conjugates. ZOO BIOLOGY 5:1-6, 1986.

de Vries, A. Translocation of mantled howling monkeys (*Alouatta palliata*) in Guanacaste, Costa Rica. M.A. thesis. University of Calgary, Alberta, Canada, 1991.

Dierenfeld, E.S.; Koontz, F.W.; Goldstein, R.S. Feed intake, digestion and passage of the proboscis monkey (*Nasalis larvatus*) in captivity. PRIMATES 33(3):399-405, 1992.

Dresser, B.L. Cryobiology, embryo transfer, and artificial insemination in *ex-situ* animal conservation programs. Pp. 296-308 in BIODIVERSITY. E.O. Wilson, ed. Washington, DC, National Academy Press, 1986.

Eudey, A. To procure or not to procure. Pp. 146-152 in ETHICS ON THE ARK: ZOOS, ANIMAL WELFARE AND WILDLIFE CONSERVATION. B.G. Norton; M. Hutchins; E.F. Stevens; T.L. Maple, eds. Washington, DC: Smithsonian Institution Press, 1995.

Gipps, J.H.W., ed. Beyond captive breeding: Reintroducing endangered mammals to the wild. SYMPOSIA OF THE ZOOLOGICAL SOCIETY OF LONDON 62:1-284, 1991.

Glander, K.E. Dispersal patterns in Costa Rican mantled howling monkeys. INTERNATIONAL JOURNAL OF PRIMATOLOGY 13(4):415-436, 1992.

Hardy, D.G. Current research activities in zoos. Pp. 531-536 in WILD MAMMALS IN CAPTIVITY: PRINCIPLES AND TECHNIQUES. D.G. Kleiman; M.E. Allen; K.V. Thompson; S. Lumpkin, eds. Chicago, University of Chicago Press, 1996.

Horwich, R.H.; Lyon, J. A BELIZEAN RAIN FOREST: THE COMMUNITY BABOON SANCTUARY. Gays Mills, Wisconsin, Orang-utan Press, 1990.

Horwich, R.H.; Koontz, F.W.; Saqui, E.; Saqui, H.; Glander, K. A reintroduction program for the conservation of the black howler monkey in Belize. ENDANGERED SPECIES UPDATE 10(6):1-6, 1993.

Hutchins, M.; Conway, W.G. Beyond Noah's Ark: The evolving role of modern zoos and aquariums in field conservation. INTERNATIONAL ZOO YEARBOOK 34:117-130, 1995.

Hutchins, M.; Wiese, R.J. Beyond genetic and demographic management: The future of the Species Survival Plan and related AZA conservation efforts. ZOO BIOLOGY 10:285-292, 1991.

Hutchins, M.; Foose, T.; Seal, U. The role of veterinary medicine in endangered species conservation. JOURNAL OF ZOO AND WILDLIFE MEDICINE 22:277-281, 1991.

Hutchins, M.; Willis, K.; Wiese, R. Strategic collection planning: theory and practice. ZOO BIOLOGY 14:5-25, 1995.

Hutchins, M.; Willis, K.; Wiese, R. Captive breeding and conservation. CONSERVATION BIOLOGY 11(1):3, 1997.

Jones, S., ed. Captive propagation and reintroduction: A strategy for preserving endangered species? ENDANGERED SPECIES UPDATE 8(1):1-88, 1990.

Karesh, W.B.; Cook, R.A. Applications of veterinary medicine to *in situ* conservation. ORYX 29(4):244-252, 1995.

Karesh, W.B.; Cook, R.A. The Wildlife Conservation Society's Field Veterinary Program. In AMERICAN ZOO AND AQUARIUM ASSOCIATION FIELD CONSERVATION MANUAL. M. Hutchins; W.G. Conway, eds. Bethesda, Maryland, American Zoo and Aquarium Association, (in press).

Kleiman, D.G. Behavioral research in zoos: Past, present and future. ZOO BIOLOGY 11:301-312, 1992.

Kleiman, D.G. Reintroduction programs. Pp. 297-305 in WILD MAMMALS IN CAPTIVITY: PRINCIPLES AND TECHNIQUES. D.G. Kleiman; M.E. Allen; K.V. Thompson; S. Lumpkin, eds. Chicago, University of Chicago Press, 1996.

Kleiman, D.G.; Allen, M.E.; Thompson, K.V.; Lumpkin, S., eds. WILD MAMMALS IN CAPTIVITY: PRINCIPLES AND TECHNIQUES. University of Chicago Press, Chicago, 1996.

Kleiman, D.G.; Beck, B.B.; Dietz, J.M.; Dietz, L.A. Costs of a reintroduction and criteria for success: Accounting and accountability in the Golden Lion Tamarin Conservation Program. Pp. 125-142 in BEYOND CAPTIVE BREEDING: REINTRODUCING ENDANGERED SPECIES TO THE WILD. J.H.W. Gipps, ed. Oxford, Clarendon Press, 1991.

Kleiman, D.G.; Beck, B.B.; Dietz, J.M.; Dietz, L.A.; Ballou, J.D.; Coimbra-Filho, A.F. Conservation program for the golden lion tamarin: Captive research and management, ecological studies, educational strategies, and reintroduction. Pp. 959-979 in PRIMATES: THE ROAD TO SELF-SUSTAINING POPULATIONS. K. Benirschke, ed. New York, Springer-Verlag, 1986.

Koebner, L. ZOO BOOK: THE EVOLUTION OF WILDLIFE CONSERVATION CENTERS. New York, Forge, 1994.

Koontz, F.W. Beam me up Scottie: Satellite tracking African elephants. WILDLIFE CONSERVATION September/October:65, 1992.

Koontz, F.W. Trading places: Reintroduction of black howler monkeys into the Cockscomb Basin, Belize. WILDLIFE CONSERVATION May/June:52-59, 1993.

Koontz, F.W., ed. AZA OLD WORLD MONKEY REGIONAL COLLECTION PLAN. Bronx, New York, Wildlife Conservation Society, 1994.

Koontz, F.W. Wild animal acquisition ethics for zoo biologists. Pp. 127-145 in ETHICS ON THE ARK: ZOOS, ANIMAL WELFARE AND WILDLIFE CONSERVATION. B.G. Norton; M. Hutchins; E.F. Stevens; T.L. Maple, eds. Washington, DC, Smithsonian Institution Press, 1995.

Koontz, F.W. Translocation of black howler monkeys (*Alouatta pigra*) in Belize: A multidisciplinary team approach for effective wildlife conservation. In AMERICAN ZOO AND AQUARIUM ASSOCIATION FIELD CONSERVATION MANUAL. M. Hutchins; W.G. Conway, eds. Bethesda, Maryland, American Zoo and Aquarium Association, (in press).

Koontz, F.W.; Horwich, R.; Saqui, E.; Glander, K. Reintroduction of the black howler monkey (*Alouatta pigra*) into the Cockscomb Basin Wildlife Sanctuary, Belize. Pp. 104-111 in PROCEEDINGS OF THE AMERICAN ASSOCIATION OF ZOOS AND AQUARIUMS ANNUAL CONFERENCE. Wheeling, West Virginia, American Zoo and Aquarium Association, 1994.

Lacy, R.C. The importance of genetic variation to the viability of mammalian populations. JOURNAL OF MAMMALOGY, (in press).

Norton, B.G.; Hutchins, M.; Stevens, E.F.; Maple, T.L., eds. ETHICS ON THE ARK: ZOOS, ANIMAL WELFARE AND WILDLIFE CONSERVATION. Washington, DC, Smithsonian Institution Press, 1995.

Oftedal, O.T.; Allen, M.E. The feeding and nutrition of omnivores with emphasis on primates. Pp. 129-138 in WILD MAMMALS IN CAPTIVITY: PRINCIPLES AND TECHNIQUES. D.G. Kleiman; M.E. Allen; K.V. Thompson; S. Lumpkin, eds. Chicago, University of Chicago Press, 1996.

Ralls, K.; Brugger, K.; Ballou, J. Inbreeding and juvenile mortality in small populations of ungulates. SCIENCE 206:1101-1103, 1979.

Ryder, O. The collection of samples for genetic analysis: Principle, protocols and pragmatism. Pp. 1033-1036 in PRIMATES: THE ROAD TO SELF-SUSTAINING POPULATIONS. K. Benirschke, ed. New York, Springer-Verlag, 1986.

Sausman, K., ed. ZOOLOGICAL PARK AND AQUARIUM FUNDAMENTALS. Wheeling, West Virginia, American Association of Zoological Parks and Aquariums, 1982.

Savage, A.; Giraldo, L. H.; Blumer, E.S.; Soto, L.H.; Burger, W.T.; Snowdon, C.T. Field techniques for monitoring cotton-top tamarin (*Saguinus oedipus oedipus*) in Colombia. AMERICAN JOURNAL OF PRIMATOLOGY 31:189-196, 1993.

Savage, A.; Giraldo, L.H.; Soto. L.H.; Snowdon, C.T. Demography, group composition and dispersal in wild cotton-top tamarins. AMERICAN JOURNAL OF PRIMATOLOGY 38:85-106, 1996.

Seal, U.S. Intensive technology in the care of *ex situ* populations of vanishing species. Pp. 289-295 in BIODIVERSITY. E.O. Wilson, ed. Washington, DC, National Academy Press, 1988.

Snyder, N.F.R.; Derrickson, S.R.; Beissinger, S.R.; Wiley, J.W.; Smith, T.B.; Toone, W.D.; Miller, B. Limitations of captive breeding programs in endangered species recovery. CONSERVATION BIOLOGY 10(2):338-348, 1996.

Stevenson, M.; Baker, A.; Foose, T.J. CONSERVATION ASSESSMENT AND MANAGEMENT PLAN FOR PRIMATES. Minneapolis, Minnesota, IUCN/Conservation Breeding Specialist Group, 1992.

Thompson-Handler, N.; Malenky, R.K.; Reinartz, G.E., eds. ACTION PLAN FOR PAN PANISCUS: REPORT ON FREE-RANGING POPULATIONS AND PROPOSALS FOR THEIR PRESERVATION. Milwaukee, Wisconsin, Zoological Society of Milwaukee County, 1995.

Wemmer, C.; Thompson, S.D. Short history of scientific research in zoological gardens. Pp. 70-94 in THE ARK EVOLVING: ZOOS AND AQUARIUMS IN TRANSITION. C. Wemmer, ed. Front Royal, Virginia, Conservation and Research Center (Smithsonian Institution), 1995.

Wemmer, C.; Rudran, R.; Dallmeier, F.; Wilson, D. Training developing country nationals is a critical ingredient to conserving global biodiversity. BIOSCIENCE 43:1-14, 1993.

Wilson, E.O. THE DIVERSITY OF LIFE. Cambridge, Massachusetts: Harvard University Press, 1989.

World Conservation Union (IUCN). THE IUCN POLICY STATEMENT OF CAPTIVE BREEDING. Gland, Switzerland, World Conservation Union, 1987.

Yeager, C.P.; Silver, S.; Dierenfeld, E.S. Phytochemical and mineral comparison of leaves from food and non-food plants of proboscis monkeys *(Nasalis larvatus)*. AMERICAN JOURNAL OF PRIMATOLOGY 41:117-128, 1997.

Dr. Fred W. Koontz is Director of the Science Resource Center at the Wildlife Conservation Society. Formerly, Dr. Koontz served as Curator of Mammals at the Bronx Zoo and as a research graduate student at the National Zoological Park, Washington, DC. Among other professional duties, he is co-chair of the American Zoo and Aquarium Association's Old World Monkey Advisory Group and a member of the IUCN's Reintroduction Specialist Group.

A Coquerel's sifaka (*Propithecus verreauxi coquereli*) relaxes at Duke University Primate Center. (Photo by David Haring.)

ZOO-BASED CONSERVATION OF MALAGASY PROSIMIANS

Sukie Zeeve and Ingrid Porton

Madagascar Fauna Group, Setauket, New York (S.Z.); St. Louis Zoo, Missouri (I.P.)

INTRODUCTION
Madagascar: Imminent Extinctions

Madagascar, the world's fourth largest island lying some 300 miles off Africa's east coast in the Indian Ocean, is widely recognized as one of the Earth's most critically imperiled centers of biodiversity. Over 90% of its animal species and 80% of its plant species are endemic, having evolved in isolation for over 150 million years since the island separated from the African mainland. Madagascar is home to an astonishing diversity of primates, with five extant endemic families comprised of 14 genera and 31 species of prosimians. Despite its relatively small size, Madagascar has the third highest number of primate species in the world, after Brazil and Indonesia [Mittermeier et al., 1994].

In the 1500-2000 years since the island was settled by humans, about 15 lemur species and most of Madagascar's other large animal species have become extinct. Many more species face imminent extinction even as they are being discovered and described. As Madagascar struggles to meet the needs of its impoverished, rapidly-growing population, its unique ecosystems are being severely degraded; over 80% of the island's vegetation has already been destroyed. Efforts to preserve Madagascar's lemurs and other unique fauna are among the conservation community's highest priorities [Zeeve, 1996]. This chapter describes the zoo community's efforts to protect the long-term survival of prosimians.

THE ROLE OF ZOOS IN CONSERVATION

The role of zoos in conservation is both multifaceted and evolving. One of the most significant contributions zoos make to the conservation of biodiversity is through education. Certainly the need to foster a conservation ethic has never been greater, as species and habitat are being lost at an accelerating rate worldwide. The opportunity to see live animals attracts large audiences and zoo education departments are developing innovative programs to enlighten these visitors about the vital role of biodiversity.

Zoos have further examined the role captive wildlife populations can contribute to the preservation of their counterparts in the wild. For certain species, that role may be as a genetic reservoir should reintroduction or restocking be determined an appropriate conservation tool. Captive populations may also serve as a fund raising focal point for *in situ* conservation efforts, as a resource for the development of technologies that can be tested and perfected prior to field application, as well as for basic scientific research. Indeed, because wild populations of many species are so small and geographically isolated they may benefit from the small population management techniques refined in zoos.

To decide which species could most benefit from captive breeding programs, the American Zoo and Aquarium Association (AZA) developed the concept of Taxon Advisory Groups (TAG). TAGs assess all species that fall under a higher taxonomic group such as a family (e.g., Felid TAG) or order (e.g., monotreme and marsupial TAGs). Using defined criteria (e.g., status in the wild, education value, research potential and needs, husbandry experience, current captive status, availability of founders), each species or subspecies is thoughtfully evaluated and ranked. Because zoo space is limited, hard decisions have to be made regarding which taxa will be recommended for captive management programs. The resulting Regional Collection Plan (RCP) is used by institutions to help select species that are part of cooperatively managed programs.

One criterion considered in the selection of species for inclusion in the RCP is whether that captive population can contribute to a conservation project in the country of origin. Helping to define opportunities or needs and to implement subsequent action within the species' range country are Fauna Interest Groups (FIGs). Composed of professionals that have experience in the country or countries encompassed by the FIG (e.g., Madagascar FIG, Southeast Asia FIG), these groups facilitate the involvement of zoos in conservation of wild populations [Wemmer & Anderson, 1992].

The Prosimian TAG

The AZA Prosimian TAG published its first RCP in 1993 and the updated and revised second edition in 1997. The second edition reflects the relationship that has evolved between the AZA TAG and its European counterpart, the EEP (European Endangered Species Programme) Prosimian TAG. By establishing cooperative breeding programs at a bi-regional level, the zoological community is working to further optimize use of limited resources, especially captive space.

The RCP reviewed the wild [Harcourt & Thornback, 1990; Mittermeier et al., 1992] and captive status [ISIS (International Species Information System, a computerized records system) studbook records, and unpublished surveys] of 31 lemur species (51 taxa). Twenty-nine taxa of lemurs are currently held in North American and/or European zoos; some are represented by only a few individuals and others by substantial populations. Employing the criteria established by the TAG, lemur taxa were placed in one of five population management categories (some of which are bi-regional programs):

1. Species Survival Plan© (SSP) - usually about 250-400 animals with a goal of maintaining 90% of the captive population's original genetic diversity for 100 years;
2. Population Management Plan (PMP) - a genetically managed population varying from 25 to 100 individuals that may require intensive husbandry research which, if successful, would allow upgrading to an SSP, or which may serve another research or education function;
3. Phase-out population - taxa maintained in captivity but which have not been recommended for SSP or PMP management and which should therefore be phased out through attrition to provide space for targeted species;
4. Insufficient Information - taxa which currently exist in North American collections but for which critical decision-making information is lacking (i.e., a breeding moratorium is requested pending further information); and
5. Not In/Do Not Bring In - taxa not held in captivity and not currently recommended for a managed program.

With input from Madagascar Fauna Group personnel and field biologists working in Madagascar, the Prosimian TAG concluded that a captive lemur population could be viewed as one component of a metapopulation management plan. Pending suitable ecological and sociological conditions, there appears to be the political will in Madagascar to participate and cooperate in well-planned reintroduction or restocking programs (such as Project Betampona, described later in this chapter). Consequently, when recommending taxa for SSPs or PMPs, the conservation status of the species was accorded significant weight. TAG recommended programs include SSPs for the ruffed lemur (*Varecia variegata variegata, V. v. rubra*), black lemur (*Eulemur macaco macaco, E. m. flavifrons*), mongoose lemur (*E. mongoz*), ringtailed lemur (*Lemur catta*), and Coquerel's mouse lemur (*Mirza coquereli*). PMP populations are recommended for the aye aye (*Daubentonia madagascariensis*), crowned lemur (*Eulemur coronatus*), collared lemur (*E. fulvus collaris*), gentle gray lemur (*Hapalemur griseus griseus*), Coquerel's sifaka (*Propithecus coquereli*), diademed sifaka (*P. diadema diadema*) and golden-crowned sifaka (*P. tattersalli*). PMP populations are also recommended for the golden bamboo lemur and giant bamboo lemur (*Hapalemur aureus* and *H. simus*), should founders become available [Porton, 1997].

The Madagascar Fauna Group

In response to international concern for lemurs, Madagascar was the first region for which a Fauna Interest Group was formed. The Madagascar Fauna Group (MFG) grew out of a 1987 workshop for conservationists and members of the Malagasy government hosted by the Wildlife Conservation Society on St. Catherine's Island. The group signed an accord, the St. Catherine's Convention for Collaboration, calling for coordinated efforts to conserve all classes of Malagasy fauna, particularly those whose populations are falling below viable minimums.

Table 1. Madagascar Fauna Group.

Steering Committee
David Anderson, Chair, San Francisco Zoological Gardens
John Behler, Wildlife Conservation Society/Bronx Zoo
John Hartley, Jersey Wildlife Preservation Trust
Jean-Marc Lernould, Parc Zoologique et Botanique de Mulhouse
George Rabb, Brookfield Zoological Park
Ulysses S. Seal, Conservation Breeding Specialist Group (ex-officio)
Elwyn Simons, Duke University Primate Center

Member Institutions
Aktiengesellschaft Zoologischer Garten Köln
The Baltimore Zoo
Brookfield Zoological Park
The Bronx Zoo (Wildlife Conservation Society)
Cincinnati Zoo & Botanical Garden
Colchester Zoo
Columbus Zoo
Denver Zoological Gardens
Duke University Primate Center
Fort Worth Zoological Park
Institute for Conservation of Tropical Environments
Jersey Wildlife Preservation Trust
Knoxville Zoological Gardens
Los Angeles Zoological Park
Marwell Preservation Trust
Micke Grove Zoo
Ogród Zoologiczny Poznan
Oklahoma City Zoological Park
Parc Zoologique et Botanique de la Ville de Mulhouse
Philadelphia Zoological Garden
Point Defiance Zoo & Aquarium
Roger Williams Park Zoo
San Antonio Zoological Gardens & Aquarium
San Francisco Zoological Gardens
St. Louis Zoological Park
University of Strasbourg/Institute Louis Pasteur
Zoo Atlanta
Zoological Society of London
Zoologischer Garten der Landeshaupstadt Saarbrücken
Zoologischer Garten Zürich

The MFG was formed in 1988 to implement the St. Catherine's agreement. Starting with nine founding institutions, membership has grown to include 30 zoos and universities in the USA, Europe and Great Britain (Table 1); members contribute annual fees to support in-country projects. Through the MFG consortium, zoos can participate in conservation projects in Madagascar even if they do not have institutional personnel or resources to implement international programs on their own.

The MFG's primary goals are improving Madagascar's two zoos; promoting field conservation and re-stocking projects; environmental education initiatives; training Malagasy nationals in wildlife management techniques; and participation of member institutions in conservation breeding, education and research. Maintaining a full-time presence in-country is a crucial initial element in implementing these goals; MFG personnel in Madagascar have established a strong working relationship with Malagasy colleagues and better coordination of projects with other conservation groups. The MFG interacts with primatologists and other field biologists from numerous institutions and serves as liaison between those individuals and governmental and non-governmental agencies working in Madagascar.

The St. Catherine's accord, renewed in 1994 for another five-year period, permits the MFG to transfer animals from Madagascar, enlisting zoos through TAGs and SSPs to assist high-priority species in need of reinforcement through captive propagation. The ability to transfer animals between in-country and out-of-country locations facilitates reintroduction or re-stocking projects, in addition to improving the genetic profiles of captive populations.

In-Country Programs for Lemur Propagation

The MFG has a primary commitment to Madagascar's two zoos: Parc Botanique et Zoologique de Tsimbazaza, in the capital city of Antananarivo, and Parc Zoologique Ivoloina, near the port city of Toamasina (Tamatave) in eastern Madagascar. Many Malagasy visitors to these parks learn for the first time about the uniqueness of Madagascar's lemurs and other native species. Some of the lemur species at these zoos are held nowhere else in the world. MFG personnel coordinate programs to increase the capacity for in-country propagation of lemurs and other taxa: (1) through programs in conservation breeding, sound animal management and conservation education; (2) by providing ongoing training for Malagasy staff; and (3) through encouraging cooperation at an international level in propagation of endangered species and their preservation in the wild.

Parc Tsimbazaza has undergone many renovations and improvements, to better reflect its mission and key role in raising conservation awareness for Malagasy visitors and students as well as for the many international tourists who visit each year. With assistance from MFG staff, general operations have improved and the animal collection is being entered into the ISIS database. Many other projects are being developed, such as: (1) review of Parc Tsimbazaza's master plan; (2) renovation of exhibits for lemurs and other taxa; (3) staff training in zoo manage-

Figure 1. The Madagascar Fauna Group is working to restock the black and white ruffed lemur (*Varecia variegata variegata*) in the Natural Reserve of Betampona. (Photo by David Haring.)

ment, record keeping, animal husbandry, English language and computer skills; (4) public education and community outreach; and (5) development of cooperative plans with other conservation groups to improve park facilities.

Parc Ivoloina, a former field station that is now Madagascar's second zoo, was developed and inaugurated with guidance from MFG Technical Advisors Andrea Katz and Charles Welch, who have overseen construction of all facilities outlined in the park's master plan. Repairs to the dam system have allowed the completion of a peninsula exhibit on a freshwater lake; the 2.5 ha peninsula exhibit has been planted with over 2500 fruit trees and forest seedlings as future food sources. The garden area has been planted with pineapples and other food items for the lemurs. The MFG funds two workers to collect seeds from trees where lemurs are known to forage; seeds are grown at Parc Ivoloina and planted on the peninsula and around the park. *Hapalemur g. griseus* and *Microcebus rufus* are free-ranging on the peninsula, and *Eulemur fulvus albifrons* and *Varecia v. variegata* are in preparation for release. The park may serve as a pre-release training site for confiscated or captive-born lemurs. The zoo provides a strong environmental education program for a growing number of school children in the Tamatave region, offers training workshops for local teachers, and presents a series of conservation displays in local communities. Katz and Welch have also developed an education and media campaign, working with Madagascar's Department of Water and Forests to address the growing number of confiscated and donated lemurs held at the park.

Through this progress, in addition to improving programs for lemurs and other endemic wildlife, both zoos are developing a higher profile and gaining more recognition and support from the Malagasy government. MFG activities help to insure that Malagasy authorities are aware of the zoos' significance to education and conservation. Parc Tsimbazaza and Parc Ivoloina are becoming major conservation centers visited by thousands of Malagasy and foreign visitors each year.

WILD LEMUR POPULATIONS: INTERVENTIONS FOR SURVIVAL

Along with assisting in-country conservation breeding programs, the MFG directs zoo-based support to rescue missions, field research, restocking efforts and habitat preservation. Below, some recent projects are described to illustrate how zoos are taking an active part in on-the-ground conservation of lemurs in Madagascar.

Project Betampona: Restocking Black and White Ruffed Lemurs

The Natural Reserve of Betampona is a 2,228 hectare protected area in eastern Madagascar near Ivoloina. The reserve, which is classified as RNI (Strict Nature Reserve), comprises one of the last remaining tracts of eastern lowland forest in a vast region where primary forest has been lost to cultivation. MFG personnel initiated several surveys to better document the status of lemur species within the reserve. It soon became clear that one species, the black and white ruffed lemur (*Varecia variegata variegata*) (Figure 1), for which there is an established captive breeding program (SSP and EEP), occurred in lower than expected numbers. More intensive surveys during the dry and rainy seasons revealed that five ruffed lemur groups totaling no more than 35 animals existed in the reserve [Welch and Katz, 1992]. Such small, isolated populations are highly vulnerable to local extinction [e.g., Gilpin and Soulé, 1986]; consequently, the MFG proceeded with further research to determine the feasibility of a restocking program using captive-born ruffed lemurs.

Beck et al. [1994] have reviewed factors that require consideration prior to proceeding with a reintroduction. A variety of biological and sociopolitical criteria should be met to improve the likelihood of success as well as to ensure that a release will not jeopardize native fauna. To that end, the MFG staff coordinated initiation of a number of studies to fill in missing information. For example, vegetation studies focusing on food resource availability in sites occupied and devoid of ruffed lemurs were carried out by a Malagasy biologist to characterize and identify areas best able to support additional lemurs. A University of Liverpool graduate student mapped ranging patterns and studied the foraging and social behavior of the resident ruffed lemur groups.

One significant concern that required resolution prior to release was the issue of black and white ruffed lemur taxonomy. The question remained whether the pelage pattern variations in the black and white form represented more than one evolutionarily significant unit. Therefore, while capturing and collaring some of the study site ruffed lemurs, hair and blood samples were obtained as part of the larger analysis of ruffed lemur systematics. The results showed no appreciable difference between black and white ruffed lemur coat patterns nor between the Betampona ruffed lemurs and the SSP population [Amato et al., unpublished data]. The research was supported in part by the MFG and by Conservation International.

An assessment of the risk of disease transmission by the reintroduced animals to the resident lemurs was carried out by the TAG and SSP Veterinary Advisor using biological samples from SSP animals, wild-born confiscated animals living at Parc Zoologique Ivoloina and four wild red ruffed lemurs [Junge and Garell, 1995]. An exhaustive literature review determined what infectious diseases have been identified in lemurs. The overall risk of disease transmission was found to be modest; however, a thorough veterinary examination protocol has been established to evaluate the health of potential release candidates.

With preliminary behavioral, ecological, genetic and medical studies completed and most of the necessary funds raised, captive bred *V. v. variegata* from the SSP population have been identified as candidates for release in the Betampona Reserve. The designated animals have been transferred to large outdoor "training facilities" at Duke University Primate Center and the Wildlife Conservation Society's St. Catherine's Wildlife Survival Center, where they will be closely monitored. The free-ranging experience should improve the animals' foraging, arboreal locomotion, and predator avoidance skills in preparation for life in the wild [e.g., Lessnau & Morland, 1995].

The *Varecia* restocking program is a priority project for the MFG, Prosimian TAG, and Ruffed Lemur SSP. Its primary goals include the development and testing of reintroduction protocols for lemurs, the integration of a captive breeding program with efforts to increase the viability of an existing but small and isolated population, as well as stimulating long-term research on the effect of habitat fragmentation on rain forest species.

Additional goals of the program include increasing public awareness, improving the conservation status of the reserve and contributing to in-country training. Indeed, two rangers trained and supported by the MFG have provided the reserve's only effective protection for several years. These rangers also assist in many of the ongoing research projects. In late 1998, ANGAP (Association National pour Gestion des Aires Protégées), the Malagasy authority for protected areas, assigned two new conservation agents to the reserve who are being trained in part by MFG staff. The increased attention and protection brought by the project will also benefit the indri, diademed sifaka, bamboo lemur, brown lemur, aye-aye, and at least four other nocturnal prosimians, as well as many bird, reptile, amphibian and invertebrate species inhabiting the reserve. Thus, the black and white ruffed lemur can appropriately be defined as a flagship species, and the true value of the program can be recognized as one method to draw attention and contribute to the preservation of a rare ecosystem.

Rescue Mission for *Hapalemur simus*

The greater bamboo lemur *Hapalemur simus* has an extremely restricted range in southeast Madagascar. About 1,000 individuals are estimated to occur in one protected area, Ranomafana National Park [Wright, 1988]. Small populations still surviving in other areas face severe threats from *tavy* (slash-and-burn agriculture), cutting of bamboo and hunting [Meier, 1987; Mittermeier et al., 1994], as well as from natural disasters such as cyclones which could devastate the remaining populations. The species' precarious status led to its receiving the "Highest

Priority Rating" by the IUCN/SSC Primate Specialist Group [Mittermeier et al., 1992] as one of the most critically endangered lemur species; the Prosimian TAG designated this lemur as a high conservation priority.

In 1995, MFG Chair David Anderson (San Francisco Zoo) investigated the status of the greater bamboo lemur in an unprotected area in southeastern Madagascar. MFG Field Advisor Patricia Wright of the Institute for Conservation of Tropical Environments (ICTE) led the expedition to a site near Karianga, where, earlier in 1995, her team had discovered that over 30 *H. simus* were compressed into a small, degraded forest fragment surrounded by farms. The bamboo lemurs at this site were seen foraging on ginger, a poor food source, as the preferred bamboo is not plentiful enough to sustain this number of animals. Five *H. simus* were captured and flown to Parc Ivoloina, where they are doing well in a large enclosure. Another expedition in June, 1996, led by MFG Steering Committee member Elwyn Simons of Duke University Primate Center, relocated three more of the remaining bamboo lemurs from the dwindling forest fragment. A pair has been sent from Parc Ivoloina to Parc Tsimbazaza, and others may be sent abroad for propagation. The goal is to form the basis of a breeding population in captivity to help protect the species from extinction.

HABITAT PROTECTION

The establishment of protected areas and reserves is of paramount importance in preserving lemur populations in the wild. Many critically threatened wildlife species occur outside of Madagascar's existing reserves, requiring new initiatives to delimit priority areas for protected status [Hannah & Hough, 1995]. Below, several recent projects describe how zoos are helping to plan and establish protected areas to benefit threatened lemur populations, in collaboration with field researchers, government authorities and other conservation organizations.

Proposed Reserve for Blue-Eyed Lemurs

A protected area for blue-eyed lemurs (*Eulemur macaco flavifrons*) (Figure 2) and other species in the Befotaka region in northwest Madagascar is closer to becoming a reality, as a result of activities of the Association pour l'Étude et la Conservation des Lémuriens, a consortium of MFG institutions Parc Zoologique et Botanique de Mulhouse, the University of Strasbourg-Institute Louis Pasteur, Aktiengesellschaft Zoologischer Garten Köln and Zoologischer Garten Saarbrücken, in collaboration with Malagasy authorities. The group has conducted preparatory studies to establish a protected reserve for the blue-eyed lemur, which occurs in a limited area in the Befotaka region, which is not protected. A study of the terrain, preceded by an aerial survey, identified zones where the forest was best conserved. An inventory of plants and animals was conducted, and other lemur species were identified in the area, including *Hapalemur griseus*, *Lepilemur*, *Microcebus*, and *Daubentonia*. Personnel from Madagascar's Department of Water and Forests (DEF) conducted the inventory with project scientists as part of the consortium's training activities. The study extended the area of the proposed

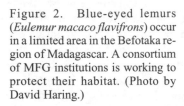

Figure 2.   Blue-eyed lemurs (*Eulemur macaco flavifrons*) occur in a limited area in the Befotaka region of Madagascar. A consortium of MFG institutions is working to protect their habitat. (Photo by David Haring.)

reserve to encompass a zone of mangrove forest and a zone of coral reef [Rumpler et al., 1995]. The project has advanced enough to be submitted to DEF in 1996 (J.-M. Lernould, personal communication).

Another of the group's studies paving the way for the proposed reserve at Befotaka utilized molecular biology techniques to compare genetic diversity in insular and mainland populations of *E. macaco*. Analysis of blood and genetic markers such as RAPDs (Random Amplified Polymorphism DNAs) or RFLPs (DNA fingerprints) showed that the *E. m. flavifrons* population of the Befotaka region present the most important reservoir of genetic variability of all the populations of *E. macaco* studied to date. The study presents genetic justification to select this region for establishing a protected reserve [Rabarivola et al., in press].

Project Masoala: Establishing a New National Park

The Masoala Peninsula in northern Madagascar encompasses the largest remaining block of lowland rain forest in the country, with a variety of unique habitat subtypes. It is home to a number of lemur species, including the red ruffed lemur (*Varecia variegata rubra*), as well as other locally endemic fauna; inventories of the peninsula's biological diversity have shown high levels of regional endemism. The Wildlife Conservation Society (WCS) is a partner in the Masoala Integrated Conservation and Development Project, a consortium of CARE International, WCS, Malagasy government agencies (ANGAP and DEF), the Peregrine Fund and Stanford Center for Conservation Biology, with funding from USAID. A major achievement was submission of a proposal in 1995, delimiting a new national park on the Masoala Peninsula; the process for official approval is now underway. The Masoala National Park will contain 210,000 ha, a buffer zone of 8,000 ha, and a forest peripheral zone of 100,000 ha, making it the largest protected area in Madagascar [Anonymous, 1995; H.S. Morland, personal communication].

The project is developing ecologically and economically sustainable activities in the forest and agricultural peripheral zones to provide alternative sources of revenue to local villagers. Harvesting schemes are being tested for primary and secondary forest products that demand forest cover, enhancing the value of intact forests to compete directly with slash and burn agriculture. The project has assisted in implementing improved agricultural methods, and in developing a marine resource management plan and systems for monitoring project impact.

The conservation component of the project, directed by Claire Kremen (WCS) and Vincent Razafimahatratra (University of Antananarivo), focused on compiling biological, socioeconomic, and forestry data; developing a program to monitor the long-term status of biodiversity and ecosystem health; and training Malagasy graduate students and project staff. For example, Adina Merenlender (Stanford University) and Marius Rakotondratsima (University of Antananarivo) studied the effects of intensive forest use by humans on two key lemur species (*Eulemur fulvus albifrons* and *Varecia variegata rubra*) and developed new methods for long-term lemur population monitoring, to develop baseline information on population composition and to track changes in group size and structure over time. Rakotondratsima received a 3-month training internship program at Stanford in 1995, and presented his work at several MFG member institutions.

Updated Census of the Lac Alaotran Gentle Lemur

Jerscy Wildlife Preservation Trust (JWPT) researcher Thomas Mutschler conducted a survey in 1995 of the Lac Alaotran gentle lemur or bandro (*Hapalemur griseus alaotrensis*), following initial work with JWPT's research officer Anna Feistner. Using transects through the marshes for censusing and radio-tracking collared animals, new data were compiled on the status of this critically endangered subspecies, which is found only in the reed beds of Lac Alaotra, Madagascar's largest lake. The survey indicated that there may be as many as 7000-12,000 bandro living among the lake's papyrus marshes, suggesting that the gene pool is still sufficient for long-term survival – if the population is sustained [Mutschler, 1995].

The report noted, however, that the bandro's range (none of which is protected) has been dramatically reduced as habitat loss and fragmentation continue at a tremendous rate. The area has one of the highest levels of human population growth in Madagascar, and Lac Alaotra is the largest rice-growing region in the country [Mittermeier et al., 1994]. Marshes in the north and south are now rice paddies, and there are official plans, supported by international aid agencies, to drain a large core area of the marsh. Timely action is critical to preserve the bandro, and urgent measures are recommended to the Malagasy government and cooperating organizations. Conservation options include establishing a protected reserve at Lac Alaotra, expanding the captive breeding program (only nine individuals are currently held worldwide), identifying alternative sites for potential bandro translocations, and intensifying the education campaign among local people.

CONCLUSIONS
The Future For Madagascar's Primates

It is all too clear that without effective intervention, the long-term survival in nature of many Malagasy prosimian species is unlikely. A comprehensive approach to lemur conservation requires the joint efforts of policy makers, field and laboratory scientists, environmental educators and conservation breeding specialists. Zoological institutions are making substantive contributions to these efforts, both in managing lemur populations in captivity and through direct participation and support for in-country projects.

The precarious status of and severity of threats to Madagascar's lemurs and other endemic wildlife have inspired a high degree of international cooperation among zoo professionals and conservation partners from numerous disciplines. These collaborative efforts have expanded our knowledge about lemurs and will help to stem the loss of species and habitat in Madagascar.

ACKNOWLEDGMENTS

Parts of this chapter were drawn from Zeeve [1996], with the kind permission of Nick Gould of the *International Zoo News*. Many thanks also to David Haring, of Duke University Primate Center, for generously providing the photographs that accompany this chapter.

REFERENCES

Anonymous. Scientists propose innovative park in Madagascar. AFRICAN WILDLIFE UPDATE 4(4):1, 1995.

Beck, B.B.; Rapport, L.G.; Price, M.S.; Wilson, A. Reintroduction of captive-born animals. Pp. 265-285 in CREATIVE CONSERVATION: INTERACTIVE MANAGEMENT OF WILD AND CAPTIVE ANIMALS. P.J.S. Olney; G.M. Mace; A. Feistner, eds. London, Chapman and Hall, 1994.

Gilpin, M.E.; Soulé, M.E. Minimum viable populations: processes of species extinction. Pp. 19-35 in CONSERVATION BIOLOGY: THE SCIENCE OF SCARCITY AND DIVERSITY. M.E. Soulé, ed. Sunderland, Massachusetts, Sinauer Associates Inc., 1986.

Hannah, L.; Hough, J., eds. GEF DESIGN ASSISTANCE TO PE2: SUMMARY OF FINDINGS. Washington, D.C., Conservation International, 1995.

Harcourt, C.; Thornback, J. LEMURS OF MADAGASCAR AND THE COMOROS. THE IUCN RED DATA BOOK. Gland, Switzerland, IUCN-The World Conservation Union, 1990.

Junge, R.E.; Garell, D. Veterinary evaluation of black and white ruffed lemurs (*Varecia variegata*) in Madagascar. PRIMATE CONSERVATION 16:44-46, 1995.

Lessnau, R.G.; Morland, H.S. Release of free-ranging black and white ruffed lemurs (*Varecia variegata variegata*) on St. Catherine's Island. AMERICAN JOURNAL PRIMATOLOGY 36(2):138, 1995.

Meier, B. Preliminary report of a field study on *Lemur rubriventer* and *Hapalemur simus* (nov. subspecies) in Ranomafana-Ifanadiana 312 Faritany Fianarantsoa, Madagascar, July 1986-January 1987. Unpublished report to Ministry of Scientific Research, Antananarivo, 1987.

Mittermeier, R.A.; Konstant, W.R.; Nicholl, M.E.; Langrand, O. LEMURS OF MADAGASCAR: AN ACTION PLAN FOR THEIR CONSERVATION, 1993-1999. Gland, Switzerland, IUCN/SSC Primate Specialist Group, 1992

Mittermeier, R.A.; Tattersall, I.; Konstant, W.R.; Meyers, D.M.; Mast, R.B. LEMURS OF MADAGASCAR. Washington, D.C., Conservation International, 1994.

Mutschler, T. Too gentle to survive? A report from the bandro field study. ON THE EDGE 72:6, 1995.

Porton, I., ed. PROSIMIAN REGIONAL COLLECTION PLAN. Second edition. St. Louis, Missouri, Prosimian Taxon Advisory Group of the American Zoo and Aquarium Association, 1997.

Rabarivola, C.; Meier, B.; Langer, C.; Scheffrahn, W.; Rumpler, Y. Population genetics of *Eulemur macaco macaco* (Primates: Lemuridae) on the islands of Nosy Be and Nosy Komba and the peninsula of Ambato (Madagascar). PRIMATES (in press).

Rumpler, Y.; Lernould, J.-M.; Nogge, G.; Ceska, V. Projet du création d'une aire protégée pour *Eulemur macaco flavifrons* au nord-ouest de Madagascar. Poster presented at the Congrés de la Ste Francaise de Primatologie, Toulouse, 1995.

Welch, C.R.; Katz, A.S. Survey and census work on lemurs in the natural reserve of Betampona in eastern Madagascar with a view to reintroductions. DODO 28:45-58, 1992.

Wemmer, C.; Anderson, D. Faunal Interest Groups: Zoo conservation with a regional focus. CBSG NEWS 3(1):14-17, 1992.

Wright, P. IUCN TROPICAL FOREST PROGRAM, CRITICAL SITES INVENTORY. Report to the World Conservation Monitoring Centre, Cambridge, U.K., 1988.

Zeeve, S. The Madagascar Fauna Group: In situ conservation initiatives for Malagasy wildlife. INTERNATIONAL ZOO NEWS 43(5):279-289, 1996.

Suzanne (Sukie) Zeeve, a former zookeeper at the Bronx Zoo, conducted doctoral research on seven primate species in the Lomako Forest, Zaire. She has worked as Biodiversity Technical Advisor at Ranomafana National Park in southeastern Madagascar and conducted research on reproductive behavior in marsupial carnivores at Point Defiance Zoo and Aquarium, Tacoma, WA. Since 1993, she has worked as Project Coordinator/USA for the Madagascar Fauna Group. **Ingrid Porton** is Mammal Curator/Primates at the St. Louis Zoological Park. She currently serves as the International Studbook Keeper and SSP Coordinator for Black Lemurs and Ruffed Lemurs and is Chair of the Prosimian Advisory Group and Vice Chair of the AZA Contraception Advisory Group.

A male cotton-top tamarin (*Saguinus oedipus*), on exhibit at Roger Williams Park Zoo, Providence Rhode Island. In 1973, the cotton-top was declared an endangered species. (Photo by Anne Savage.)

DEVELOPING A CONSERVATION ACTION PROGRAM FOR THE COTTON-TOP TAMARIN *(SAGUINUS OEDIPUS)*

Anne Savage, Humberto Giraldo, and Luis Soto
Roger Williams Park Zoo, Providence, Rhode Island (A.S.);
Proyecto Tití, Colosó, Colombia (H.G., L.S.)

INTRODUCTION

Today's modern zoos go far beyond captive breeding of threatened and endangered species [Norton et al., 1995]. As conservation resource centers of the future [Koontz, this volume], zoos now support field and education programs, provide the public with much needed information on the world's dwindling wildlife and habitat, and offer educational opportunities for individuals to experience and develop a life-long appreciation for wildlife and the need to conserve resources.

At the Roger Williams Park Zoo our commitment to conservation is highlighted through various collaborative programs such as the American Zoo and Aquarium Association's Species Survival Plans© (SSPs), effective conservation education programs on-site, nationally, and internationally [Savage, 1993; 1996a; Savage & Giraldo, 1993; Savage et al., 1989; in press, a], and through the Sophie Danforth Conservation Biology Fund small grants program [see Vecchio, in press, for a complete review]. Using our limited financial resources, we have carefully chosen the type of programs that compliment our mission to protect endangered species and their habitat.

To highlight our strong commitment to species and habitat preservation, Roger Williams Park Zoo has joined in the effort to preserve the endangered cotton-top tamarin *(Saguinus oedipus)* in Colombia. By collaborating with institutions both nationally and internationally, as well as serving as the host institution for the Cotton-top Tamarin SSP [Savage, 1996b], our zoo can make a significant contribution to the conservation of this endangered species, as well as serving as a model for other small zoos in an attempt to encourage multi-institutional conservation efforts. In this chapter, we provide details of past, current, and planned projects comprising a comprehensive conservation action program for the cotton-top tamarin.

HISTORY

The cotton-top tamarin, found only in Colombia, is one of the most endangered primates in the world. During the late 1960s and early 1970s, between 20,000-30,000 cotton-top tamarins were exported to the U.S. for use in biomedical research [Hernández-Camacho & Cooper, 1976]. In 1973, the species was declared endangered and importation was banned. Cotton-top tamarins have been kept in zoos and laboratories since the early 1960s. Today, the captive population numbers more than 1,800 individuals, with sixty-four percent of the population found in research laboratories [Tardif & Colley, 1988]. The cotton-top tamarin is the only nonhuman primate that spontaneously develops colonic adenocarcinoma and has been the primary model to study the ontogeny of this disease for many years [see Clapp, 1993, for a complete review]. The remaining captive population is found in zoos or in private institutions. In an effort to manage the current captive population, an International Studbook for the Cotton-top Tamarin, maintained by Dr. William Langbauer of the Pittsburgh Zoo, is currently being updated. The Cotton-top Tamarin SSP actively manages the more than 220 individuals found in 55 zoos in the U.S. and Canada [Savage, 1995, 1996b].

DEVELOPING A CONSERVATION PROGRAM FOR THE COTTON-TOP TAMARIN

Developing effective long-term conservation programs for endangered species requires more than just captive breeding and scientific studies. We suggest that successful conservation programs should use a multidisciplinary approach that combines field research and effective scientific assessment of habitats, as well as community initiatives that involve local inhabitants in culturally relevant, action-based programs. Making the conservation of natural habitats and resources economically feasible for local communities will help insure success. It is also important that the information derived from field-based conservation programs be communicated to and experienced by zoo visitors to increase their interest, knowledge, and support of conservation programs.

Our goals in the development of a conservation program for the cotton-top tamarin were to: (1) conduct long-term field studies to evaluate the wild population, as well as providing information from which to compare previous captive studies; (2) promote community awareness and public education programs in Colombia; (3) establish training programs in conservation biology; (4) provide economic incentives for habitat preservation; (5) develop material that could be used in zoos and public education programs to promote the conservation of the cotton-top tamarin and its habitat; and (6) foster an international collaboration between scientists, conservationists, and educators from zoos, universities and conservation organizations.

PROYECTO TITÍ

Proyecto Tití is a comprehensive, collaborative program between scientists, conservationists and educators from zoos, universities and conservation organizations in the U.S. and in Colombia to protect cotton-top tamarins and their habitat. In 1987, a collaborative agreement between the University of Wisconsin-Madison and INDERENA (now the Ministerio de Ambiente), the flora and fauna protection agency of Colombia, established an effective program for conserving cotton-top tamarins in Colombia. In 1990, Roger Williams Park Zoo and Fossil Rim Foundation joined the team to provide additional resources and field expertise and, in 1993, staff from The Wilds assisted in our program.

FIELD STUDIES: THE REPRODUCTIVE BIOLOGY OF
THE COTTON-TOP TAMARIN

This small New World primate is found only in the tropical forests of the northwest region of Colombia [Mast et al., 1993]. Prior to our studies, very little information was available on free-ranging cotton-top tamarins, with Neyman [1977] providing the only published field study on this species. Our goal was to examine some of the critical issues of callitrichid biology, patterns of reproduction, infant development and social structure. Our study began in 1988, in the Montes de Maria Reserve in Colosó, Colombia. We have developed techniques that have allowed us to identify individual animals and locate groups with ease [Savage et al., 1993], enabling us to collect detailed observations on our study groups.

We have collected data on group composition, stability, birth seasons, and dispersal patterns to examine the reproductive strategies and tactics used by male and female cotton-top tamarins [Savage et al., 1996a]. We have observed both monogamous groups and groups containing more than one reproductively active female. All of our study groups contained at least one adult female and male, with several groups containing several adult males and females. Both males and females dispersed to neighboring groups, and we have not found any differences in rates of emigrations. Males were more likely to immigrate into a new group following the death/emigration of a resident male. A review of the reproductive strategies used by cotton-top tamarins can be found in Savage et al. [1996a].

In most stable, captive groups of cotton-top tamarins have only one reproductively active female and females (daughters) that are housed in the natal group are reproductively suppressed [Savage et al., 1988; Ziegler et al., 1987b; Widowski et al., 1990]. Most captive females conceive on their first postpartum ovulation approximately 18 days after parturition [Ziegler et al., 1987a], although Ziegler et al. [1990] have also found that this interval can vary depending on the number of offspring surviving and the pattern of nursing behavior. In contrast, not all wild females are reproductively suppressed while living in the natal group [Savage et al., in review]. Unlike captive females that are capable of conceiving 18-days postpartum, wild cotton-top tamarin females show no ovarian activity for approximately 120 days postpartum. Studies investigating what controls this suppression of fertility in the wild are ongoing [see Savage et al., in review, for more detail].

The patterns of reproduction in wild cotton-top tamarins differ from their captive counterparts [Savage et al., in review]. Wild tamarins generally give birth to twins once a year. In contrast captive females can either give birth every 28 weeks [Snowdon et al., 1985] or exhibit a bimodal distribution of interbirth intervals [Kirkwood et al., 1985]. Fertility in the wild can be affected by environmental changes, yet, when environmental conditions are appropriate, females will produce live offspring that have a high probability of surviving.

The factor that appears to be most important in influencing infant survivorship in wild cotton-top tamarins is group size. Large groups of tamarins were more successful at rearing offspring than smaller groups. Not only do large groups have more individuals to assist in carrying infants, thereby reducing the energetic costs of carrying, but they are more likely to increase their foraging efficiency and predator detection. Similar results have been found with captive cotton-top tamarins. For additional information on parental care patterns and development of infant cotton-top tamarins see Savage et al. (1996b).

We are also collaborating on studies that investigate the genetic diversity of this population, in an attempt to enhance our long-term management of this species in the wild (C. Faulkes, personal communication). We have participated in a study examining Mhc class I and II genes in cotton-top tamarins. The considerable polymorphism and allelic diversity of the Mhc class I and II genes are thought to increase the number of different peptides bound and recognized by the immune system. Hughes and Nei [1988] have suggested that this polymorphism may facilitate the recognition of many different pathogens, thereby conferring a selective advantage to individuals or populations. While many species appear to be have a highly polymorphic Mch loci, captive cotton-top tamarins exhibit limited Mhc class I polymorphism [Watkins et al., 1988]. It could not be determined whether the limited Mch class I polymorphism was the result of inbreeding in captive tamarin populations or due to a sampling error resulting from a small pool of potential founders. By analyzing Mhc class I and II genes in wild cotton-top tamarins, results indicated that wild tamarins exhibited low levels of polymorphism and allelic diversity at the major histocompatability complex Mhc class I and II loci [Gyllensten et al., 1994]. Thus, Gyllensten et al. [1994] and Watkins et al. [1988] suggested that the low levels of Mch class I polymorphism and allelic diversity may potentially limit the tamarins' immune system to respond to pathogens, rendering the species more susceptible to infections.

The information generated from both the field and captive studies has significant value as long-term conservation programs are developed for this species. Data from the field studies are currently used in modeling programs (U. Seal, personal communication) to examine the long-term viability of this population in the wild, as well as to compare and validate hypotheses generated from captive studies. Most importantly, the scientific information is critically important for use by Colombian officials as they develop regional and national plans and priorities for species and habitat conservation.

PROYECTO TITÍ AND PUBLIC AWARENESS CAMPAIGNS

For conservation education efforts to be effective in Colombia, support and interest must be forthcoming from the local population. Without community support, the efforts of research scientists and conservationists are in vain. In 1988, we conducted a survey of the local school children near our study site to assess the communities' perception of the conservation needs of the area [Savage et al., 1989]. We found that many students had a variety of myths and misconceptions about the forest and the wildlife of the area. Approximately 70% of the high school students had never visited the forest even though it is only 4 km away from their village. Another disturbing fact was that over 90% of the students had no idea that the cotton-top tamarin was endemic to Colombia and not found in other countries in South America.

To increase public awareness and create an interest in conserving cotton-tops, we developed several community programs for the local villages. We distributed T-shirts produced by Conservation International and posters of cotton-top tamarins created and produced by Jersey Wildlife Preservation Trust and Penscynor Wildlife Park. These posters showed cotton-top tamarins living in their natural environment, surrounded by the plants and animals found in the forest. The posters had a positive effect in bringing cotton-top tamarins into the lives of the local people, particularly the children. Since the T-shirts were worn by members of their family and posters were displayed in their homes, young children often questioned their parents who shared information about cotton-top tamarins and their habitat. Now most children grow up with a keen sense of pride and enthusiasm in protecting this endangered species and its habitat.

As support for our program grew, we obtained a small grant from the Captive Breeding Specialist Group and matching funds from a local paint supplier in Colombia, so the children of the village could "advertise" conservation to all that passed by. Two of the elementary schools are located on the main street in the village. Each of the schools have cement walls or sheet metal walls facing the street. These walls were divided into sections so that each class would have their own space. The children, teachers, and parents of each school worked on a group project depicting scenes from the forest, the future of the forest, and the new logo for Colosó that discouraged using sling shots to capture or hunt animals. The project used older students as "mentors" to assist in the actual drawing of the scenes and the younger children to paint the scenes. The entire village became involved and local television covered the event, bringing even greater attention to this community's commitment to conservation.

Although protected by the CITES and the Endangered Species Act, cotton-top tamarins are still commonly found in Colombia's pet trade [Mast & Cubberly, 1987; Mast & Patiño, 1988]. In an effort to direct public attention to illegal capture of wildlife for the pet trade, we developed a program modeled after the successful trading of guns for money, toys, etc., in the U.S. We encouraged villagers to trade their sling shots, commonly used to hunt and capture animals for the pet trade, for stuffed cotton-top tamarin toys. Since toys are a prized possession for most young people in the region, we were very successful in generating support for this program. But most important, the hunting of wildlife for the pet trade has significantly decreased in the region (B. Avila, personal communication).

CONSERVATION EDUCATION

Building on local support, we developed several classroom and field activities that have been very successful in increasing student awareness and interest in local conservation activities [Savage et al., 1993]. Our program aims to reach all students in the local village and activities are designed to meet the needs of elementary, junior and senior high school students. In collaboration with teachers from the village, the Junta Ecológica (Colosó's local environmental group), and staff of Proyecto Tití, we developed a comprehensive program that is still used in the school system to date. Based on material developed by the education department of the Roger Williams Park Zoo, we constructed a basic package of material that focused on rainforest conservation for elementary school children. This package included material that teachers can incorporate into classroom lectures and activities for the children. We have also distributed this material to neighboring villages.

High school students participated in a hands-on field biology training program designed to study a newly formed group of cotton-top tamarins in a nearby village. The students received weekly lectures on tamarin behavior and biology and spent 8 hours each week watching the tamarins. The students, equipped with their "Proyecto Tití" T-shirts, binoculars, and notebooks were successfully trained as field biologists to monitor the animals. Their scores on the final exam of the training program indicated that they had developed an understanding of the biology and behavior of the cotton-top tamarin and their willingness and enthusiasm to continually monitor the animals revealed their dedication to the project.

These same students were also taken to the forest where they observed the animals that are part of our long-term field study. The students used the learned observation techniques to follow the animals in their natural habitat and developed an appreciation for the forest that was being rapidly destroyed. Their interest in the forest and the animals continued to grow so much that we decided to use these high school students in a peer-teaching program. Each year, elementary school children participate in a field trip to the forest. The instructors of this program are not the local biologists, they are the high school students from Proyecto Tití. The students share their knowledge of the animals and the importance of preserving the forest with the younger students. This type of peer-teaching program is exactly what is needed to make conservation a reality in any local community. If the local villagers take pride in the forest, as these students have, only then will the conservation programs developed by biologists and conservation organizations become effective.

One of the most encouraging aspects of this intensive training are the future benefits to the community. Several graduates of the field program have become elementary school teachers. With this interest and new-found knowledge of species and habitat conservation, these teachers have been extremely motivated in engaging their students in activities that will impact long-term conservation efforts in the area.

Of the thirty-eight students who have participated in our field biology training program, we have hired three former graduates of this program to assist us in various new projects. This not only made our conservation program more manageable, but allowed our team to grow and include more local support. Our goal is to increase the number of participants in this program and expand to neighboring communities.

INTERNATIONAL CONSERVATION EDUCATION PROGRAMS

Using the three phases that inspire creative action in children (awareness, affinity and action), we developed a program that used the cotton-top tamarin as a "flagship species" for conservation, but highlighted the importance of understanding culture, the conservation problem, and exploring practical solutions in developing a conservation plan. Junior high school students in Colombia and Rhode Island initiated an international letter and video exchange aimed at increasing awareness of conservation issues that impact on the long-term survival of the cotton-top tamarin (e.g., water pollution, erosion, habitat destruction, species preservation) [Savage et al., in press, a]. Tom Sullivan, an independent producer who has worked with us on this project since 1989, instructed students and teachers in the use of the video equipment and basic filming techniques. Students completed introductory videos highlighting their communities and the conservation programs. This international exchange of information created quite an impact on students wanting to learn about other cultures and conservation programs.

To expand our program, we have focused on assessing the quality of the natural resources in Colombia. In 1992, northern Colombia experienced a severe drought. Water was scarce and a dramatic increase in mortality of both domestic and wild animals was observed. Channeling community concern, we developed a program to monitor the long-term health of the watersheds in the village. Students in Colosó adopted a monitoring site near their school. Each month, students conducted standardized tests to examine the quality of the water consumed by the communities. Using the data they collected, they began to consider ways to improve the quality of the water they consumed. Students learned that protecting watersheds helps to improve the long-term health and survival of the forests, plant and animal populations, and ultimately human communities.

We have been awarded several U.S. Environmental Protection Agency grants to expand our international exchange program. Using middle school teens and teachers, we have a collaborative program that monitors the quality of our water sources in the U.S. and in Colombia. Students and teachers have the opportunity to investigate what is happening "in their own backyards," as well as address some of the growing global concerns of the long-term conservation of our natural resources.

April 6, 1996, was an exciting day for the participants of our international exchange program, as the Roger Williams Park Zoo hosted the first annual "Waters of Our World" conference. Students from the U.S. and representatives from our program in Colombia shared their results and action plans and also examined water conservation issues from a global resource perspective. Participants examined the factors critical to preserving watersheds, as well as practical solutions for saving forest habitats. Issues relating to growing human populations and deforestation were also highlighted as students participated in an Oxfam lunch, cooked using *bindes* (see below). To promote stewardship, students have designed T-shirts with practical choices that can be incorporated into our daily lives. Such choices can have a profound impact on conserving resources in Colombia and in the U.S and empowering students to take action. A future "Waters of the World" conference is planned for the students in Colombia.

COLOMBIAN STAFF/STUDENT DEVELOPMENT

We are committed to providing training opportunities for our Colombian research team and students whenever possible. The staff of Proyecto Tití have been involved in various training programs in captive management, field biology, and conservation education in Colombia and other countries. This continued investment in our staff has helped us to develop confident, well-informed leaders of the community.

We also realize that the staff of Proyecto Tití are, in many ways, unique and can serve as a model for others. Thus, whenever possible, we provide training in field biology, field veterinary medicine techniques and grass roots conservation methods for students and conservationists from Colombia and other Latin American countries. Moreover, we are deeply committed to disseminating scientific information in Colombia, through various publications or lectures at Colombian universities.

DEVELOPING ALTERNATIVES TO FOREST DESTRUCTION

Given the dramatic rate of forest destruction for human and agricultural consumption [Myers, 1984], it is critical that programs are developed to reduce the dependency on non-sustainable forest products if long-term efforts to conserve habitat for cotton-top tamarins are to be effective. The community of Colosó is located in one of the most economically depressed departments of Colombia. Although there is electricity in the village, the majority of the population cooks over an open fire. This is due to the high cost of electricity, in addition to the cultural norm of preparing food with a smoky flavor. Thus, we began a study to investigate how human harvesting of trees for firewood was affecting the long-term survival of cotton-top tamarin habitat [Savage et al., in press, b]. We surveyed 100 families and found that a family of five individuals consumes 15 ± 3.2 logs of wood (1-1.5 m in length) daily. Given this high rate of consumption and without efforts to replenish the trees that are harvested, the forested regions of Colosó faced a substantial yearly loss.

To decrease forest consumption, we examined the feasibility of using solar box cookers. This has been promoted as a viable alternative since food is cooked by solar energy. We conducted a study in which five families were instructed in the use of the solar ovens and asked to evaluate their effectiveness. There was an overwhelming negative response to the solar oven for several reasons: (1) coffee could not be made in the oven; (2) food cooked in the oven did not have an appealing taste. Even with the addition of products which mimicked the flavor of food cooked over an open fire, the food prepared in solar box cookers was unsatisfactory; (3) because of the lengthy cooking time, solar box cookers were only useful for preparing dinner; and (4) it was difficult to reheat food quickly.

Taking the criticisms of solar box cookers into consideration, we examined another traditional method of cooking. Some inhabitants of these communities use *bindes* to cook their food. Villagers collect large termite mounds from the forest, bring them back to their homes, and reinforce them with mud. These *bindes* have a hole cut at the top, yet are still strong enough to support the weight of a large kettle, and a hole cut in the side so that wood can be fed directly into a fire. Smaller holes are cut on the top and side which allow sufficient air exchange to support a fire. Villagers have told us that *bindes* are much more efficient in burning wood and they produce less smoke, which has been implicated in several women's health issues in the community. Despite the numerous benefits, the use

Figure 1. This clay *binde* was designed as a more fuel-efficient, quick, and healthy way to cook food in rural Colombia. (Photo by Anne Savage.)

Table 1. Firewood Consumption Using *Bindes* or Three Stones

	Fuel	Quantity	Average Time to Cook *Sanchocho*[1]
3 Stones (N=10)	Firewood	5950 g (15±3.2 logs)	1 hr 15 min
***Bindes* (N=10)**	Firewood	1675 g (5±2.7 logs)	1 hr 10 min
Bindes (N=10)	Yucca stalks	9100 g	2 hr 30 min
Bindes (N=10)	Corn cobs	3200 g	1 hr 29 min

[1] *Sanchoco* is a traditional meal for this region of Colombia.

of termite mounds was problematic. It is quite labor intensive to search the forest for the termite mound and, in addition, they cannot withstand constant daily use. On average, a traditional *binde* may last as long as one month with constant use.

Given that *bindes* were already culturally acceptable, we were interested in modifying the materials of a *binde* that would allow for greater long-term use. With funding provided by the Disney Foundation Conservation Excellence Fund, we began one of our most successful programs. After consulting several sources, we contacted a local artisan who creates small items out of clay. Using his expertise, a prototype clay *binde* was designed and tested (Figure 1). The community of Colosó, was invited to participate in a demonstration of the effectiveness and versatility of the newly designed *binde*. Several salient features emerged from this new prototype: (1) refuse such as corn cobs, corn husks, coconut shells, etc., could be burned just as efficiently as wood; (2) food could be cooked as quickly as using the three stone method; and (3) significantly less smoke was produced which is likely to result in less hazard for women's health. Forty families participated in a comparative study examining the effectiveness and efficiency of using the traditional method of cooking over three stones or a clay *binde*. Our study concluded that *bindes* were significantly more efficient, burning 2/3 less wood per day than cooking over three stones (Table 1). Food cooked using a clay *binde* retained its flavor and women reported less eye and lung irritation from the smoke. The demand for clay *bindes* that cost approximately $3-4 U.S. to produce has been tremendous. Efforts to expand the program to neighboring communities and develop sustainable resource areas are under investigation.

Figure 2. A male cotton-top tamarin (*Saguinus oedipus*) wears a battery-operated radio transmitter, while living in "The Tamarin Trail" - a free-ranging exhibit at the Roger Williams Park Zoo. (Photo by Anne Savage.)

LOCAL IMPACT IN COLOMBIA

Community involvement and dedication to conservation continues to climb steadily in this small village. Recently, funds have been allocated in the local budget for reforestation efforts and conservation education projects. We are pleased that our program is now reaching beyond our nearby village of Colosó. Neighboring villages are interested in using our material and becoming part of the developing team to make conservation a priority in this region. Thus, our program continues to teach students and adults important concepts about conservation and motivates them to take action and initiate changes. We feel that continued development and support of this program will not only insure the survival of the cotton-top tamarin, but will make conservation a priority for the future generations of Colombians in this region.

THE RHODE ISLAND CONNECTION

Actively involving our visitors to the Roger Williams Park Zoo in field conservation efforts is a challenging task. However, to give our visitors a chance to experience what it is like to study tamarins in a forest, we created "The Tamarin Trail." This exhibit featured a group of tamarins free-ranging in a densely wooded area of the zoo (Figure 2). Modeled after the golden lion tamarin (*Leontopithecus rosalia*) release program at the National Zoo [Bronikowski et al., 1989; Stoinski et al., this volume], visitors had the opportunity to experience "a day in the life of a field biologist." This exhibit allowed the public to view these primates in conditions which are similar to those in the wild. In addition, visitors saw first hand what it is like to study wild monkeys using the technology that is currently employed in primate field studies. In addition to numerous volunteers manning interpretive stations and various graphics describing the program, we wanted to encourage our visitors to take action and think about what they could do to conserve cotton-top tamarins and their habitat. Public interest in this exhibit was tremendous as evidenced by an increase in attendance and media attention. Expanding on this interest, the Roger Williams Park Zoo has committed to developing an interactive exhibit that will address the issues facing the long-term conservation of this species in Colombia.

CONCLUSIONS

The cotton-top tamarin is an endangered primate from northwestern Colombia. Efforts to conserve this species in the wild and in captivity have resulted in the development of an integrated approach to conserve habitat and increase public awareness and action in local communities. Through the efforts of *in situ* and *ex situ* programs highlighting the cotton-top tamarin as the flagship species for conservation in Colombia, significant progress has been made. This program can serve as a model for zoos and other conservation organizations in its collaborative nature, attention to culturally appropriate programs, economically based alternatives, and long-term commitment to species and habitat conservation.

We continue to seek ways to expand our team of scientists and educators to incorporate new ideas and concepts both locally and internationally. We are committed to bringing international attention to the plight of the cotton-top tamarin, yet focusing on making local action in conservation a reality in Colombia. We feel that continued development and support of this program will not only insure the survival of the cotton-top tamarin, but will make conservation a priority for the future generations of Colombians in this region.

ACKNOWLEDGMENTS

Support for our programs in Colombia is made possible by grants from the National Science Foundation, U.S. Aid for International Development, U.S. Environmental Protection Agency, Institute of Museum Services, The Disney Foundation for Conservation Excellence, American Society of Primatologists, Rhode Island Zoological Society, Roger Williams Park Zoo (RWPZ), AAZK Chapter of the RWPZ, RWPZ Docent Council and various individual contributions. We wish to thank the staff of Proyecto Tití, Dr. A. Villa and Dr. B. Avila for their continued support of our work in Colombia. We would like to recognize the significant contributions of C. Snowdon, E. Blumer and T. Sullivan in the success of this program.

REFERENCES

Bronikowski, E.J.; Beck, B.B.; Power, M. Innovation, exhibition and conservation: Free-ranging tamarins at the National Zoological Park. Pp. 540-546 in PROCEEDINGS OF THE AMERICAN ASSOCIATION OF ZOOLOGICAL PARKS AND AQUARIUMS ANNUAL CONFERENCE. Wheeling, West Virginia, American Zoological and Aquarium Association, 1989.

Clapp, N.K., ed. A PRIMATE MODEL FOR THE STUDY OF COLITIS AND COLONIC CARCINOMA, THE COTTON-TOP TAMARIN *SAGUINUS OEDIPUS*. United States of America, CRC Press, Inc., 1993.

Gyllensten, U.; Bergstrom, T.; Josefsson, A.; Sundvall , M.; Savage, A.; Giraldo, L.H.; Blumer, E.S.; Watkins, D.I. The cotton-top tamarin revisited: Limited Mhc Class I polymorphism of wild tamarins and limited nucleotide diversity of the class II DQA1, DQB1 and DRB Loci. IMMUNOGENETICS 40(3):167-176, 1994.

Hernández-Camacho, J.; Cooper, R.W. The nonhuman primates of Colombia. Pp. 35-69 in NEOTROPICAL PRIMATES: FIELD STUDIES AND CONSERVATION. R.W. Thorington; P.G. Heltne, eds. Washington, D.C., National Academy of Sciences, 1976.

Hughes, A.L.; Nei, M. Pattern of nucleotides substitution at major histocompatibility complex class I loci reveals overdominant selection. NATURE 335:167-170, 1988.

Kirkwood, J.K.; Epstein, M.A.; Terlecki, A.J.; Underwood, S.J. Breeding a second generation of cotton-top tamarins (*Saguinus oedipus oedipus*) in captivity. LABORATORY ANIMALS 19:269-272, 1985.

Mast, R.; Cubberly, P.S. SOS for the cotton-top tamarin. FOCUS 9(2):5, 1987.

Mast, R.B.; Patiño, A.F. Aid for a native Colombia. NATURE CONSERVANCY MAGAZINE, January/February, 1988.

Mast, R.B; Rodriguez, J.V.; Mittermeier, R.A. The Colombian cotton-top tamarin in the wild. Pp. 3-44 in A PRIMATE MODEL FOR THE STUDY OF COLITIS AND COLONIC CARCINOMA, THE COTTON-TOP TAMARIN *SAGUINUS OEDIPUS*. N.K. Clapp, ed. United States of America, CRC Press, Inc., 1993

Myers, N. THE PRIMARY SOURCE, TROPICAL FORESTS AND OUR FUTURE. New York, W.W. Norton & Company, Inc., 1984.

Neyman, P.E. Aspects of the ecology and social organization of free-ranging cotton-top tamarins (*Saguinus oedipus*) and the conservation status of the species. Pp. 39-67 in THE BIOLOGY AND CONSERVATION OF THE CALLITRICHIDAE. D.G. Kleiman, ed. Washington, D.C., Smithsonian Institution Press, 1977.

Norton, B.G.; Hutchins, M.; Stevens, E.F.; Maple, T.L., eds. ETHICS ON THE ARK: ZOOS, ANIMAL WELFARE, AND WILDLIFE CONSERVATION. Washington, D.C., Smithsonian Institution Press, 1995.

Savage, A. Tamarins, teens and teamwork: An integrated approach to in situ conservation. Pp. 654-658 in PROCEEDINGS OF THE AMERICAN ASSOCIATION OF ZOOLOGICAL PARKS AND AQUARIUMS REGIONAL CONFERENCE. Wheeling, West Virginia, American Zoological and Aquarium Association, 1993.

Savage, A. COTTON-TOP TAMARIN (*SAGUINUS OEDIPUS*) AZA SSP© MASTERPLAN, Providence, Rhode Island, Roger Williams Park Zoo, 1995.

Savage, A. AZA Species Survival Plan Profile: The cotton-top tamarin - Managing populations in captivity and creating community concern in Colombia. ENDANGERED SPECIES UPDATE 13(3):9-11, 1996a.

Savage, A. 1996 COTTON-TOP TAMARIN (*SAGUINUS OEDIPUS*) AZA SSP© MASTERPLAN. Providence, Rhode Island, Roger Williams Park Zoo, 1996b.

Savage, A.; Giraldo, H. "Proyecto Tití:" The development of a conservation education program in Colombia. FIRST PAN AMERICAN CONFERENCE ON THE CONSERVATION OF WILDLIFE THROUGH EDUCATION PROCEEDINGS, 1993.

Savage, A.; Giraldo, L.H.; Blumer, E.S.; Soto, L.H.; Burger, W.; Snowdon, C.T. Field techniques for monitoring cotton-top tamarins (*Saguinus oedipus oedipus*) in Colombia. AMERICAN JOURNAL OF PRIMATOLOGY 31:189-196, 1993.

Savage, A.; Giraldo, L.H.; Soto, L.H. Proyecto Tití: Developing global support for local conservation. AZA FIELD CONSERVATION RESOURCE GUIDE. M. Hutchins; W. Conway, eds. (in press, a).

Savage, A.; Giraldo, L.H.; Soto, L.H.; Snowdon, C.T. Demography, group composition and dispersal in wild cotton-top tamarins. AMERICAN JOURNAL OF PRIMATOLOGY 38(1):85-100, 1996a.

Savage, A.; Shideler, S.E.; Soto, L.H.; Causado, J.; Lasely, B.L.; Snowdon, C.T. Monitoring reproductive cycles of wild cotton-top tamarins females (*Saguinus oedipus*) using fecal steroid analysis. AMERICAN JOURNAL OF PRIMATOLOGY, in review.

Savage, A.; Snowdon, C.T.; Giraldo, H. Proyecto Tití: A hands-on approach to conservation education in Colombia. Pp. 323-326 in PROCEEDINGS OF THE AMERICAN ASSOCIATION OF ZOOLOGICAL PARKS AND AQUARIUMS ANNUAL CONFERENCE. Wheeling, West Virginia, American Zoological and Aquarium Association, 1989.

Savage, A.; Snowdon, C.T.; Giraldo, H.L.; Soto, L.H. Parental care patterns and vigilance in wild cotton-top tamarins (*Saguinus oedipus*). Pp. 187-199 in ADAPTIVE RADIATIONS OF NEOTROPICAL PRIMATES. M. Norconk; A. Rosenberger; P. Garber, eds., New York, Plenum Press, 1996b.

Savage, A.; Soto, L.H.; Giraldo, L.H. Proyecto Tití: Developing alternatives to forest destruction. In: PRIMATE CONSERVATION - A RETROSPECTIVE AND A LOOK AT THE 21ST CENTURY. R.A. Mittermeier; A. Rylands, eds. (in press, b).

Savage, A.; Ziegler, T.E.; Snowdon, C.T. Sociosexual development, pair bond formation, and mechanisms of fertility suppression in female cotton-top tamarins (*Saguinus oedipus oedipus*). AMERICAN JOURNAL OF PRIMATOLOGY 14:345-359, 1988.

Snowdon, C.T.; Savage, A.; McConnell, P.B. A breeding colony of cotton-top tamarins (*Saguinus oedipus oedipus*). LABORATORY ANIMAL SCIENCE 35:477-480, 1985.

Tardif, S.D.; Colley, R. INTERNATIONAL COTTON-TOP TAMARIN STUDBOOK. Oak Ridge, Tennessee, Oak Ridge Associated Universities, 1988.

Vecchio, A.J. The Sophie Danforth Conservation Biology Fund of the Roger Williams Park Zoo. AZA FIELD CONSERVATION RESOURCE GUIDE. M. Hutchins; W. Conway, eds. (in press).

Watkins, D.I.; Hodi, F.S.; Letvin, N.L. A primate species with limited major histocompatibility complex class I variability. PROCEEDINGS OF THE NATIONAL ACADEMY OF SCIENCE USA 85:7714-7718, 1988.

Widowski, T.M.; Ziegler, T.E.; Elowson, A.M.; Snowdon, C.T. The role of males in the stimulation of reproductive function in female cotton-top tamarins, *Saguinus o. oedipus*. ANIMAL BEHAVIOR 40:731-740, 1990.

Ziegler, T.E.; Bridson, W.E.; Snowdon, C.T.; Eman, S. Urinary gonadotropin and estrogen excretion during the postpartum estrus, conception and pregnancy in the cotton-top tamarin (*Saguinus oedipus oedipus*). AMERICAN JOURNAL OF PRIMATOLOGY 12:127-140, 1987a.

Ziegler, T.E.; Savage, A.; Scheffler, G.; Snowdon, C.T. The endocrinology of puberty and reproductive functioning in female cotton-top tamarins (*Saguinus oedipus oedipus*) under varying social conditions. BIOLOGY OF REPRODUCTION 36: 327-342, 1987b.

Ziegler, T.E.; Snowdon, C.T.; Uno, H. Social interactions and determinants of ovulation in tamarins (*Saguinus*). Pp. 113-133 in SOCIOENDOCRINOLOGY OF PRIMATE REPRODUCTION. T.E. Ziegler; F.B. Bercovitch, eds. New York, Wiley-Liss, Inc., 1990.

Dr. Anne Savage is the Director of Research for the Roger Williams Park Zoo and Adjunct Assistant Professor of Population Biology at Brown University. She is the SSP Coordinator for the Cotton-Top Tamarin and is a member of the AZA's New World Primate Taxon Advisory Group and Board of Regents. **Luis Soto** is the Field Project Coordinator and **Humberto Giraldo** is the Staff Biologist for Proyecto Tití. In 1994, the Roger Williams Park Zoo received the AZA's Significant Achievement in Conservation Award for Proyecto Tití.

A group of golden lion tamarins (*Leontopithecus rosalia*), on exhibit at the National Zoo in Washington, DC. (Photo by Jessie Cohen.)

THE GATEWAY ZOO PROGRAM:
A RECENT INITIATIVE IN GOLDEN LION TAMARIN
REINTRODUCTIONS

Tara Stoinski, Benjamin Beck, Mary Bowman, and John Lehnhardt

Zoo Atlanta, Atlanta, Georgia (T.S, M.B.), National Zoo, Washington, D.C. (B.B., J.L.), and Georgia Institute of Technology, Georgia (T.S.)

INTRODUCTION

The drastic loss of habitats and species witnessed in the twentieth century has spawned an increased interest in reintroducing captive-bred animals into the wild. Zoological parks, state and federal wildlife agencies, universities and independent individuals have all become active in reintroduction programs. However, reintroductions, regardless of their sponsors, are often unsuccessful. According to Kleiman [1989], 50% of the estimated 1000 Asian attempts have failed, and only four or five of twenty mammalian reintroductions have established viable populations. Beck et al. [1994] found similar results – only 11% of the 145 captive reintroduction programs examined by the authors have been successful. The reasons for such depressing statistics are to a large extent still unknown, as many programs fail to document and quantify their results; Beck et al. [1994] found that two years of intensive searching produced information on post-release outcomes for less than 50% of known reintroductions.

The Golden Lion Tamarin Conservation Program (GLTCP) represents one of the most recent mammalian reintroduction attempts and one of the few efforts to reintroduce a captive-born primate species. Since 1984, reintroduction has played a central role in the GLTCP and has been successful in increasing the number of golden lion tamarins living in the wild. Over the past 12 years, the reintroduction methodology of the GLTCP has changed, and recently a new initiative aimed at creating more uniform and predictable reintroductions, educating the public, and improving reintroduction techniques has been developed. Termed the Gateway Zoo Program, this effort is now utilized by the GLTCP to help ensure the future of the golden lion tamarin in the wild. This chapter provides a brief history of the GLTCP, including a description of the innovative free-ranging golden lion tamarin zoo exhibits. In addition, we provide complete details of the Gateway Zoo Program and the methods used to predict reintroduction success.

BACKGROUND ON THE GOLDEN LION TAMARIN
CONSERVATION PROGRAM

Golden lion tamarins (*Leontopithecus rosalia*) are New World primates endemic to the Atlantic Coastal Rainforest of Brazil. The species is a member of a unique family of primates, the callitrichidae, which includes the tamarins and marmosets. Callitrichids are characterized by their small size, cooperative infant care, and production of twin offspring (Figure 1). Weighing less than two pounds, golden lion tamarins are omnivores, feeding on fruits, insects, and small vertebrates. They live in small groups consisting of one pair of breeding adults plus younger animals, who are often the adults' offspring. Groups actively defend territories of approximately 100 acres from other golden lion tamarins through vocalizations and scent marking.

Large-scale deforestation in Brazil over the last three centuries destroyed much of the golden lion tamarins' traditional habitat. This, coupled with poaching for the pet trade, resulted in a drastic decrease in the wild population; by the 1960s the golden lion tamarin was declared one of the world's most critically endangered primate species [Kleiman et al., 1986]. Concurrent with the decline in wild numbers was a crisis among the captive population. In 1972, an international survey found that only 70 tamarins were maintained in captivity, and extinction appeared imminent in this population as mortality rates exceeded birth rates [Kleiman et al., 1986]. The survey results combined with the situation in the wild precipitated an international conference that same year to address the fate of the golden lion tamarin and other callitrichid species. One of the first initiatives to develop from the conference was an intensive research effort aimed at investigating the reproduction, social behavior and husbandry of captive golden lion tamarins. Headed by Dr. Devra Kleiman at the National Zoological Park, the studies found that tama-

Figure 1.   An adult male golden lion tamarin relaxes with his twin offspring. (Photo by Joe Sebo.)

rins were being inappropriately fed and housed. Changes in diet and social housing conditions resulted in a tremendous increase in golden lion tamarin numbers during the 1970s and 1980s, such that by 1984 the captive population numbered 371 animals [Kleiman et al., 1991]. Zoos currently house over 500 golden lion tamarins and these individuals are managed internationally to ensure the future of the captive population.

The increase in captive numbers in the 1970s and 1980s was accompanied by substantial improvements in conservation efforts in Brazil. By 1973, a captive breeding facility for endangered primates endemic to Brazil had been created in Rio de Janeiro. Additionally, the Poco das Antas Reserve was established in 1974 to protect the remaining golden lion tamarin habitats. The fusion of efforts in Brazil with those at National Zoo was the basis for the Golden Lion Tamarin Conservation Program (GLTCP). Initially established in response to the threats facing the captive and wild populations, the GLTCP has evolved into an international, multidisciplinary conservation effort aimed at saving the Atlantic Coastal Rainforest through the preservation of a flagship species, the golden lion tamarin [Golden Lion Tamarin Conservation Program, 1991].

Reintroduction: A Brief History

With the emergence of a self-sustaining captive population and the establishment of the Poco das Antas Reserve, the reintroduction of captive-bred golden lion tamarins into Brazil became possible in the early 1980s. GLTCP researchers viewed reintroduction as a beneficial methodology for increasing the size, genetic diversity, and geographic distribution of the wild population. Thus, in 1983, field studies were initiated to provide baseline behavioral and ecological data on wild tamarins and to locate appropriate release sites. Captive animals were then selected for reintroduction based on genetics; individuals that were genetically over-represented or that could mate with other reintroductees without inbreeding were favored [Beck et al., unpublished data]. After selection, tamarins were exposed to a pre-release training protocol, designed specifically to promote and strengthen natural foraging skills. Training was initiated at National Zoo and continued at the Rio de Janeiro Primate Center where the animals were quarantined for six months after shipment to Brazil [Beck et al., 1991].

Two to four weeks before release, the tamarins were moved from the primate center to large acclimatization cages built around natural vegetation at the release site. Formal training continued in these cages, and potential prey and predators provided additional experiences for the tamarins [Beck et al., 1991]. Between May and July, 1984, 14 animals, comprising three pairs and one family group, were released into the wild [Kleiman et al., 1986]. Despite the intensive pre-release training, the reintroduced animals showed striking deficits in locomotion and orientation, and only 5 of the 14 released were still free-ranging in the reserve eight months later. By June, 1985, 11 of the released animals had died or had been removed [Beck et al., unpublished data].

Figure 2. A golden lion tamarin (*Leontopithecus rosalia*) wearing a radio collar while in a free-ranging pre-release training program at the National Zoo. (Photo by Joe Sebo.)

In response to the 1984 results, the reintroduction protocol was altered for 1985. Researchers redesigned pre-release training to specifically address the locomotor and orientation deficits observed in 1984. In addition, researchers assessed the effects of pre-release training by training only one of the two groups slated for release. Data collected after reintroduction on ranging, foraging effort, and foraging success quickly demonstrated that untrained tamarins performed worse than trained individuals. Training of the naive group was then initiated while the animals were loose. This post-release training was feasible because the animals remained psychologically tethered to the release site, i.e., they were not motivated to leave the food and nestbox area. The training resulted in an increase in foraging, locomotion and navigation skills in just a few weeks and, by the end of the first year, no discernible differences existed between the two groups [Beck et al., unpublished data].

Two important conclusions were drawn from the 1985 reintroductions: training is necessary but not sufficient for survival, and training can occur after release. The success of the post-release training utilized in 1985, combined with the high costs and limited benefits of pre-release training in 1984, led the GLTCP scientists to consider post-release training as a new reintroduction. Researchers believed that a tamarin family scheduled for reintroduction could be quarantined at a donor zoo, shipped to a release site, and then immediately released [Bronikowski et al., 1989]. Thus, in 1986 an experiment was conducted at the National Zoo to explore this possibility. A family of tamarins was quarantined at a donor institution and shipped to National Zoo where they underwent the same set of procedures that precede a reintroduction; the individuals were weighed, dye-marked and fitted with radio collars (Figure 2). However, instead of being released in Brazil, the tamarins were freed in a remote area of National Zoo which was supplemented with feeding devices to stimulate foraging, and ropes and vines to mimic a natural rainforest habitat. Initially, the animals were reluctant to stray far from the nestbox, but over a two month period, foraging, locomotor and orientation competence increased. Thus, this post-release training provided the animals with an opportunity to learn general skills needed for survival, a procedure that was less time- and labor-intensive than the previous training of specific skills [Bronikowski et al., 1989].

The success of the experiment convinced GLTCP scientists that general skills training conducted after release was an effective strategy for preparing captive tamarins for reintroduction. As a result, reintroductions in 1987 utilized only post-release training, which was conducted in Brazil. Tamarins were shipped from donor institutions via National Zoo to Brazil and were released within 10 days of arrival, after which training was conducted. As with previous reintroductions, extensive post-release support occurred during this time: the animals were fed daily, returned to shelter boxes when they became disoriented, and treated when injured. Researchers observed that this post-release training methodology, in addition to streamlining reintroduction, increased the survival rate of the reintroduced tamarins [Bronikowski et al., 1989]. Consequently, the same reintroduction protocol was used in subsequent years but with one alteration – each tamarin group also spent one to three months free-ranging at National Zoo before being shipped to Brazil. Thus, the tamarins were exposed to pre-release training in the zoo and post-release training in Brazil. This new procedure offered several benefits; the tamarins were provided with two opportunities to learn general skills necessary for survival; the combination of pre- and post-release training was the most cost-effective reintroduction methodology used thus far; and having free-ranging tamarins in the zoo created an opportunity to educate zoo visitors about the plight of golden lion tamarins and the work of the GLTCP.

To maximize the education opportunities for the public, the free-ranging tamarins became part of a zoo exhibit in 1988. A release site accessible to visitors was selected and educational messages were provided through informative graphics and informational talks led by trained volunteers. The exhibit was very popular with the public, as the presence of monkeys in the trees delighted zoo visitors and encouraged an active dialogue between visitors and volunteers [Bronikowski et al., 1989]. The success of the exhibit, coupled with the cost-effectiveness of the pre- and post-release training methodology, resulted in the permanent incorporation of the free-ranging period into the reintroduction protocol. The current reintroduction procedure now requires all captive tamarins slated for release to have free-ranging experience in a zoo before being shipped to Brazil.

In summary, reintroduction has played an integral role in the conservation of golden lion tamarins since 1984. Because reintroduction is a newly emerging science, it has operated as a dynamic process within the GLTCP, continually evolving over the last 12 years in response to new findings. For example, early reintroductions clearly demonstrated that training protocols designed to teach specific skills, such as foraging or locomotion, did not significantly affect post-release survival. As a result, a general skills training procotol was subsequently initiated and, since 1988, the combination of pre-release training in the zoo combined with post-release training in the field has been the methodology used to teach captive-bred tamarins general skills necessary for successful reintroduction. Such changes have helped produce a substantial increase in the number of golden lion tamarins in the wild. As of December 1995, the reintroduced population equaled 169 animals and represented 23% of the wild population living at liberty in the Atlantic Coastal Rainforest [Beck & Martins, 1995].

Golden lion tamarin reintroductions have also improved our understanding of this conservation procedure and have provided a valuable database for current and future reintroduction attempts. For example, the finding that specific skills training does not translate into post-release success has been replicated in other reintroductions. Vargas [1994] reported that, while black-footed ferrets provided with predatory training were more proficient at killing live prairie dogs than were untrained individuals, this superior ability did not translate into enhanced post-release survival. Similar results have been found with Siberian polecats [Miller et al., 1994], and studies focusing on specific skills training in a variety of species have shown limited lasting results of such training [Miller et al., 1990; I. McLean et al., unpublished data]. These results suggest that other variables, such as environmental experience, may play a central role in reintroduction success; future research will address such hypotheses.

Despite the advances made by golden lion tamarin reintroductions, the GLTCP is still far from its ultimate goal of 2000 self-sustaining tamarins. GLTCP scientists feel that a self-sustaining population may be achieved without further captive additions due to the extensive breeding within the reintroduced population and the high survivorship of these wild-born offspring. However, Beck and Martins [1995] point out that reintroduction is still necessary to provide genetic diversity in the wild population, improve the genetic and demographic status of the captive population, promote conservation education, and maintain support for the program by the zoo community. Thus, reintroduction will continue to function as a GLTCP strategy and new ways of improving the overall process and success rate will continue to be sought. One current GLTCP project facilitating this process is the Gateway Zoo Program.

THE GATEWAY ZOO PROGRAM

The success of the free-ranging program at National Zoo in the late 1980s established a need for more free-ranging exhibits to train individuals slated for release. Thus, in the 1990s the GLTCP began involving additional zoos, and the Gateway Zoo Program was born. Over the past five years, five institutions have joined National Zoo in the gateway program: Zoo Atlanta (Georgia), Brookfield

Zoo (Chicago, Illinois), Los Angeles Zoo (California), Metro Washington Park Zoo (Portland, Oregon), and Milwaukee Zoo (Wisconsin). In order to qualify as a gateway zoo, these institutions were required to submit to the GLTCP an application form, information questionnaire, three letters of recommendation, and a letter regarding implementation of the free-ranging exhibit. Additionally, potential gateway zoos were required to have previous experience breeding callitrichids and were required to sign a cooperative research and management agreement developed to establish guidelines for the management of the captive population.

With the expansion of the reintroduction protocol to include gateway zoos, a set of standard rules was required to ensure uniform and streamlined reintroduction preparation among all institutions. Drafted by GLTCP scientists, these rules are still in effect today. First, gateway zoos are required to house a single reproductive group for up to two years to provide extensive pre-release training experience and ample opportunity to reproduce; successful reproduction permits the shipment of entire family groups to Brazil, which is more cost-effective and more likely to result in survivorship than the shipment of pairs. Second, gateway zoos are responsible for the quarantine of animals, providing medical care and handling shipping requirements and costs. Third, each zoo is required to provide training opportunities through the free-ranging program and is encouraged to participate in a standardized data collection procedure developed by the GLTCP.

Quarantine and Medical Requirements

A critical factor in reintroducing captive tamarins is avoiding the introduction of infectious diseases or genetic abnormalities into the wild population. Therefore, strict medical and genetic screening of reintroduction candidates has occurred throughout the history of the GLTCP. With the creation of the Gateway Zoo Program, the GLTCP developed standardized quarantine and screening protocols for both gateway institutions and donor zoos supplying the animals. Standard quarantine procedures for the donor zoo include at least two weeks of quarantine in a facility which allows no direct or indirect contact with other primates and which provides a separate air supply and keeper staff. Standard medical screening includes: (a) fecal exams to test for endoparasites, (b) body weights and oral exam, (c) cultures for pathogens, such as *Salmonella*, (d) hematology and serum chemistry screening, and (e) callitrichid hepatitis screening. If the latter three tests are not part of the donor zoo's capabilities, they can be performed at the gateway institution. Standard genetic screening consists of a diaphragmatic hernia test; this condition is common among captive golden lion tamarins and may affect success in the wild. Any abnormal results are reported to the GLTCP's veterinary staff, and animals that do not qualify for reintroduction are replaced.

After completion of quarantine and screening at the donor zoo, animals are shipped to the gateway institution and quarantined for at least 30 days. Medical screening at the gateway zoo consists of two fecal exams, representing samples from three consecutive days, and cultures for pathogenic bacteria. Additionally, any tests not performed at the donor institution are also conducted at this time. Abnormalities are reported to the GLTCP's veterinary staff, and animals that fail this final health exam are returned to the captive population.

Shipping and Permit Requirements

Numerous sets of shipping and permit requirements are needed to get each tamarin from its original institution to Brazil. In order to streamline the transfer of animals between donor and gateway zoos, shipping and permit requirements are arranged by direct contact between the institutions involved. Permits necessary for shipping are acquired by the gateway institution and the donor zoo pays the cost of transport. Gateway zoos are responsible for all shipping costs to Brazil, which includes the modification of transport kennels to meet international regulations. Import and export permits required by CITES are obtained by the National Zoo. Both donor and gateway institutions are responsible for providing National Zoo with any information needed for these permits.

The Free-Ranging Exhibit

A third requirement of gateway zoos is to provide the animals with pre-release training through a free-ranging exhibit. While the different terrain, vegetation types, and weather conditions of the zoos make each free-ranging exhibit unique, gateway institutions follow a set of basic rules established by the GLTCP:

Figure 3. Golden lion tamarin with a tail dye-mark which makes him easy to identify at a long distance. (Photo by Jessie Cohen.)

1. The exhibit area is separated from any obvious dangers and must provide suitable sub-canopy pathways for travel [Bronikowski et al., 1989]. Sites lacking in understory are often improved by the addition of ropes and vines.
2. To prevent losses, all tamarins are fitted with a radio collar and remain under continuous human observation while they are outside the nestbox.
3. Tamarins receive tail dye-marks for easy identification (Figure 3).
4. Plastic coolers are mounted in trees as nestboxes to provide protective shelter and an easy method of capture.
5. Feeding platforms are suspended approximately five meters off the ground. These platforms force animals to travel to obtain food and are purposely unsturdy to mimic natural substrates. Additional devices, such as cricket feeders, are also used to increase foraging, and supplemental food is distributed throughout the exhibit to encourage navigation and manipulation of the environment with the hands (micromanipulation).
6. Educational information for the public is provided through informative graphics and informational talks led by trained volunteers.

The free-ranging exhibit is more labor intensive than most traditional zoo exhibits, which require little human involvement outside of keeper tasks, due to the lack of separation between the free-ranging tamarins and the rest of the zoo. For example, the tamarins must remain under constant supervision while they are free-ranging to minimize harm or loss. Additionally, the public must be prevented from interacting directly with the animals. These extra needs are met through the use of volunteers. Under the supervision of a coordinator, a group of trained volunteers is responsible for fulfilling the various requirements of the free-ranging exhibit, such as maintaining contact with the animals and providing information to the public. Volunteers are taught necessary skills and given a background on golden lion tamarins and the GLTCP at a mandatory training class developed by their gateway institution. Because of the numerous responsibilities of the free-ranging exhibit, many gateway zoos require that two or more volunteers be present, especially during peak visitation times. Using multiple volunteers ensures that both tamarin watching and exhibit interpretation can occur simultaneously.

Research Protocol and Data Collection in the Free-Ranging Exhibit

There have been various forms of data collection during the ten years of the free-ranging program. In 1986, the primary goal was research and National Zoo volunteers collected behavioral data throughout the free-ranging period. Between 1987 and 1989, the focus of the free-ranging program changed from research to exhibitry and, as a result, minimal data collection occurred to allow volunteers to concentrate on public education. With the advent of the Gateway Zoo Program, GLTCP scientists recommended that all institutions collect location information on each tamarin. This protocol has been utilized by most gateway zoos for the last five years. In 1995, GLTCP scientists began to pilot a new research protocol aimed at developing a methodology for predicting post-release success from free-ranging behavior and biographical information. It is hoped that this research will lead to explosive growth in the reintroduced population. Improving the survival rate of reintroduced animals increases both their number and the probability that these individuals will have an opportunity to reproduce. Wild-born offspring are highly successful and thus an accelerated increase in their numbers is likely. Additionally, the ability to predict which animals will survive will add to our understanding of reintroduction as a science.

Predicting Post-Release Success: Methods. Two main criteria guided the development of the new protocol. The data collected during the free-ranging period needed to be sufficiently similar to data collected after release in Brazil to facilitate comparison. In addition, the protocol needed to be easily learned and executed by volunteers. After a review of the Brazilian protocol by the first two authors, modifications were made which resulted in the use of both scan and continuous sampling methodologies. Instantaneous scans were conducted every 15 minutes to gather information on five categories of information relevant to post-release success: behavior, height off the ground, substrate type and diameter, activity, and location. Continuous data were collected on three rare behaviors: falls, near falls, and descents to the ground. In 1995 this protocol was piloted at National Zoo and Zoo Atlanta with the intention that, if the data collection was reliable, the research program would be expanded to include all gateway zoos.

While the same data were collected at both institutions, subdivisions in two of the scan categories (behavior and height) varied slightly between the two locations due to the fact that National Zoo originally expected to include five animals and thus wanted a slightly easier protocol. For the behavior category, Beck selected five behaviors from the Brazilian protocol that were relevant to reintroduction success: eat/drink, micromanipulation, rest, other and social. National Zoo utilized these categories plus a 'not visible' category for behavioral observations. Zoo Atlanta used the same ethogram with two changes. First, instead of lumping social behaviors into one category, seven individual social behaviors were listed with an 'other social' category to capture any additional social behaviors. Collapsing these eight behaviors produced National Zoo's social behavior category. Second, an additional two behaviors found on the Brazilian protocol were included: autogroom and coprophagy. Comparisons between institutions will focus only on the six behaviors found on the National Zoo ethogram. Divisions in the height category also differed for the two institutions. Height at Zoo Atlanta was divided into three meter increments up to 15.5 meters. National Zoo divided height into one meter increments up to 4.5 meters; heights above 4.5 meters were not categorized. Collapsing National Zoo's height data into three meter increments will permit comparison of the data.

Divisions in the substrate, activity and location categories were the same for both zoos. Six types of substrates were recorded: (a) natural substrates with a diameter less than 2 cm, (b) natural substrates with a diameter between 2 cm and 5 cm, (c) natural substrates with a diameter greater than 5 cm, (d) man-made substrates, (e) the ground, and (f) in the nestbox. Activity was divided into stationary, locomotor, and 'not visible' categories. Finally, location data were collected by dividing each exhibit area into 10 m by 10 m grids outlined with yellow nylon rope secured to the ground.

Observations were recorded using pencil and paper. Data collection shifts were two hours in length, beginning at 0700 hours and ending at 1900 hours or whenever the tamarins entered the nestbox for the night. At the beginning of each shift, information on weather conditions and temperature was recorded. A stopwatch was used to signal scan data collection times. The first scan data were collected 15 minutes after the start of a shift to prevent volunteers from randomly initiating data collection in response to the animals' behavior.

Volunteers at both institutions were trained in general research methodology and the specifics of the golden lion tamarin protocol. The majority of Zoo Atlanta's volunteers had no previous research experience, and thus the first stage of training concentrated on providing basic information on data collection in the zoo setting. Led by Zoo Atlanta's lead research volunteer, this discussion touched on how, why, and when research is conducted, and identified the relationship of research to Zoo Atlanta's mission. During the second stage of training, the research coordinator led volunteers through each step of the data collection process, from operating the equipment to recording data. Volunteers were instructed to collect scan data by taking a mental snapshot of the animals at the time of the scan and recording relevant data from this snapshot. If the tamarins were separated during a scan, volunteers were told to record data on the visible animal first and then on the non-visible animal when it was located. If this animal was not located within five minutes, it was listed as 'not visible' for the scan.

Toward the end of the training session, volunteers collected data from both still pictures and a video. The results of each 'scan' were discussed at length and any questions were answered. Finally, volunteers were encouraged to practice collecting data on both the permanent exhibit tamarins and the free-ranging tamarins once they were released. The training was required of each research volunteer and, in addition, a reliability test, which involved the volunteer collecting data simultaneously with one of the program coordinators, was required before data collection was permitted.

A slightly different training procedure was utilized at National Zoo. At a single group session led by Beck, volunteers were taught the ethogram and the above described methodology for collecting data. During the first several weeks that the animals were free-ranging, volunteers practiced collecting data with the assistance of keepers, research department staff, the volunteer coordinator, and other research volunteers. Unlike Zoo Atlanta, approximately one-third of National Zoo's volunteers had previous research experience and thus were able to assist with training. Data were considered valid when the volunteer coordinator had collected data with an individual and assessed him or her as reliable.

In addition to the free-ranging data, a biographical profile is being compiled on each of the free-ranging animals. Through surveys distributed to each previous housing institution, we are collecting information on the husbandry practices, housing conditions, and social experiences to which the animals have been exposed. This information will be combined with the free-ranging behavior data and then utilized in developing the prediction model.

Results. The amount of data collected over the course of the 1995 summer was substantial. At National Zoo, 75 volunteers collected approximately 1360 hours of data between 19 June and 16 October. Data were collected an average of 12 hours per day, which equaled the entire time the animals were free-ranging. At Zoo Atlanta, 46 volunteers collected 736 hours of data between 14 July and 21 October. Data were collected an average of eight hours a day and had an average reliability of approximately 90%.

Final results of the data collection will not be available until after the animals are released in Brazil. The current plan is to store the free-ranging data plus information gained from the biographical surveys until one or two years after release. At that time, the data bank will be analyzed for life-history and behavior factors that correlate with post-release survival. The information gained will be incorporated into a methodology for predicting the post-release success of individual tamarins from their early histories and behavior at gateway zoos.

A preliminary look at the data reveals interesting differences between individual tamarins. For example, Zoo Atlanta's male was observed on natural substrates less than 2 cm in diameter in 31.8% of scans while the female was observed on this substrate category in only 22.2% of scans. Additionally, National Zoo's male was observed on natural substrates in 25.5% of scans while the female was observed on natural substrates in only 15.3% of scans. Large differences between the tamarins were also observed in allogrooming behavior (Zoo Atlanta), height utilization (Zoo Atlanta) and substrate use (National Zoo). These individual differences suggest that pre-release training is unique for each animal. If differential survival among the reintroduced tamarins correlates with the individual behavioral differences observed, we will have identified possible predictors of post-release success.

The data also reveal differences between institutions. For example, both Zoo Atlanta animals were observed on natural substrates more often than National Zoo tamarins (24.2% of scans versus 10.2% of scans). Additionally, both National Zoo animals were not visible (42.4% of scans) more often than the Zoo Atlanta tamarins (11.8% of scans). Other differences between institutions include the amount of micromanipulation observed and the percentage of scans where animals were stationary or locomoting. The similarities in these areas between individuals at the same institution suggest that many of these differences may be due to exhibit design. For example, the Zoo Atlanta exhibit contains several stands of bamboo, and traveling in these areas requires the use of very thin substrates. The exhibit at National Zoo, however, is located in a stand of mature oak, holly, and beech trees. As a result, animals are able to successfully locomote without having to use thin substrates.

The data also provide a method for reviewing the effectiveness of the free-ranging period. By looking at the frequency of occurrence over time, the progress of the tamarins in each of the scan categories can be assessed and any problems with the exhibit can be modified. For example, at both zoos the average frequency of natural substrate use decreased over time while the frequency of human-made substrate use increased over time. Such results are disappointing in that animals are expected to increase their use of natural substrates over the course of free-ranging. However, they provide us with valuable information for the next free-ranging session. For example, one potential reason for the observed results at Zoo Atlanta is that supplemental food was often placed on fences or cages. As a result, the animals appeared to increase both the amount of time spent on such substrates and their frequency of visits. In response to these results, the 1996 exhibit will be modified such that supplemental food will be placed only on natural substrates. The results at National Zoo can be accounted for by a large platform in the middle of the exhibit which became a favorite resting place of the tamarins; this platform will be removed prior to the next free-ranging season.

Research conclusions. The 1995 research protocol proved successful in many ways. In addition to the eventual use of the free-ranging data in predicting post-release success, this information allows us to track the progress of the tamarins over the course of the summer and institute changes to the exhibit. Furthermore, the presence of data collectors intrigued zoo visitors, enabling more opportunities for public education.

In addition to highlighting the potential for a successful research program, our pilot study also illuminated some possible problems. First, differences in volunteer training techniques at the two institutions could result in possible differences in data collection techniques. Although teaching styles were different, extensive conversations between individuals at Zoo Atlanta and National Zoo were conducted to minimize any differences in the teaching of the actual methodology. Second, no inter-rater reliability was conducted between the two zoos. Logistically, it would be difficult – but not impossible – to accomplish this goal, and it may be something the research program moves toward as more gateway institutions participate in the data collection protocol. Third, the differences in behavior observed between institutions suggest that drawing broad conclusions based on golden lion tamarins at different zoos may be problematic. While various differences do exist between exhibits, attempts are made to provide similar locomotor and foraging experiences at all gateway zoos by supplementing existing vegetation and providing foraging opportunities. Free-ranging tamarins are thus provided with a similar framework from which training evolves. Additionally, because free-ranging is aimed at teaching general skills as opposed to specific skills, differences in exhibits are not of crucial importance. For example, general orientation skills can be effectively learned in either a patch of mature oak or a stand of bamboo. Differences between free-ranging exhibits would be more problematic if training involved specific skills that depend on the physical environment of the exhibit. However, it will be interesting to see if any of the observed differences correlate with post-release success.

As a result of the success in 1995, the research program will be expanded in 1996. First, Zoo Atlanta and National Zoo will collect data on their returning pairs. Second, additional gateway zoos will participate in data collection; Metro Washington Park Zoo and Los Angeles Zoo are in the process of reviewing the protocol and will begin training volunteers in the spring. Third, additional tamarin pairs will be established for the first of two seasons of free-ranging at the remaining gateway zoos, and it is hoped that data collection will occur at these institutions as well. In future years, the project will continue to evolve in response to the results of the data collection and attempts will be made to standardize data collection between gateway institutions.

GATEWAY ZOOS, REINTRODUCTION AND THE GLTCP

What is the impact of the Gateway Zoo Program on the goals of reintroduction and the GLTCP? Since the first tamarins were reintroduced into Brazil in 1984, reintroduction has been aimed at achieving six objectives: (a) increasing the size of the wild population, (b) increasing the genetic diversity of the wild population, (c) expanding the geographic distribution of the wild population, (d) protecting additional tracts of Atlantic Coastal Rainforest, (e) contributing to the science of reintroduction, and (f) enhancing programs of public education [Beck et al., 1991]. The impact of the Gateway Zoo Program on the these objectives is substantial. First, gateway zoos provide the most cost-effective and successful method thus far for preparing captive golden lion tamarins for reintroduction. Second, gateway zoos educate the public through free-ranging exhibits. Third, the Gateway Zoo Program has mobilized support for reintroduction in the zoo community. Finally, by conducting research, gateway zoos are contributing to the science of reintroduction and improving success for future reintroductions.

As a part of the GLTCP's integrated conservation effort, reintroduction represents just one of the methodologies used by the GLTCP to achieve its objectives. As stated earlier, the GLTCP originated in the early 1970s in response to the threat of extinction of both the wild and captive golden lion tamarin populations. While the GLTCP's original objectives were to increase the number of captive individuals and protect remaining habitat in Brazil, the scope and impact of the project has expanded considerably in the last 25 years. Thus, current GLTCP objectives include: (a) maximizing the probability of survival of the wild golden lion tamarin population, (b) expanding the techniques used in conservation biology, (c) increasing public awareness and involvement in the conservation of golden lion tamarins and their habitat, (d) enhancing profession training in conservation, and (e) increasing the impact of conservation through integration with other similarly purposed programs [Golden Lion Tamarin Conservation Program, 1991].

As a methodology of the GLTCP, reintroduction has impacted each of the above objectives over the last 12 years. For example, reintroduction has helped maximize the probability of survival of the wild golden lion tamarin population by increasing numbers and genetic diversity; it has expanded techniques used in conservation biology by continually re-evaluating reintroduction techniques; it has increased public awareness and involvement by providing jobs for local inhabitants and encouraging local farmers to protect golden lion tamarins on their land; it has enhanced professional training in conservation by providing research opportunities for students; and it has increased the impact of conservation by providing data and methodologies for newly emerging reintroduction attempts. The importance of reintroduction to GLTCP objectives highlights the need for initiatives like the Gateway Zoo Program to further our understanding of this science.

While the focus of this chapter has been on reintroduction and the gateway zoo, it is important to remember that these programs represent only a small portion of the GLTCP. In addition to reintroducing captive-bred animals, the program focuses on habitat preservation and restoration, global management of the captive population, demographic and behavioral biology studies of the wild population, studies of the flora and fauna of the Atlantic Coastal Rainforest and conservation education [Golden Lion Tamarin Conservation Program, 1991]. These methodologies contribute significantly to the overall impact of the GLTCP and play an important role in reintroduction success. For example, habitat preservation and restoration has resulted in a 45% increase in the amount of protected area available for tamarins, thus expanding the area available for reintroductions. Additionally, conservation education has helped decrease the amount of loss due to human theft and vandalism. This interdependence means that the success of the GLTCP is determined by the sum of its parts; no one methodology or objective within the GLTCP is more important than any other. It is this holistic approach to conservation that is responsible for the success of the GLTCP to date.

CONCLUSIONS

Since the 1980s, the reintroduction of captive-bred animals has played a central role in the GLTCP long-term plan and, since 1991, the gateway zoos have been integral in preparing animals for release in Brazil. However, the impact of the Gateway Zoo Program extends beyond pre-release training. It is a methodology for standardizing reintroduction procedures, educating the public and providing scientific training. Additionally, information gained from the current gateway zoo research will further our understanding of reintroduction as a science and increase the numbers in reintroduced populations. These benefits, combined with the continued importance of reintroduction to the GLTCP, ensure that the Gateway Zoo Program will play a significant role in future reintroductions.

ACKNOWLEDGMENTS

Zoo Atlanta and National Zoological Park would like to thank the numerous staff and volunteers who cared for and observed the free-ranging tamarins. The Golden Lion Tamarin Conservation Program is supported by the National Zoological Park and has been funded by grants from the Smithsonian Institution (International Environmental Sciences Program and Scholarly Studies); Friends of the National Zoo; World Wildlife Fund for Nature (WWF-US); Frankfurt Zoological Society-Help for Threatened Wildlife, Germany; National Science Foundation (grant #SBR9318900); Jersey Wildlife Preservation Trust, UK; Lion Tamarins of Brazil Fund; National Geographic Society; Wildlife Preservation Trust International; TransBrasil Airline, Brazil; Fundacão O Boticario de Protecão a Natureza, Brazil; IUCN; Calgary Zoo and Botanical Garden, Canada; The Lincoln Park Zoological Gardens; Philadelphia Zoological Society; Chicago Zoological Society; National Council for Research, Brazil; Harezo Shimizu, Brazil; Japan Marmoset Institute, Japan; Whitley Wildlife Trust, UK; William P. McClure; The British Embassy, Brazil; and the Canadian Embassy, Brazil; European Union; Federal Republic of Germany; Rain Forest Trust Fund; and the Ministry of Environment of Brazil.

Collaborating institutions include: The Smithsonian Institute; Associacão Mico Leão Dourado, Brazil; National Council for Research, Brazil; Instituto Brasileiro do Meio Ambiente e dos Recursos Naturais Renovaveis (IBAMA), Brazil; Centro de Primatologia de Rio de Janeiro (CPRJ-FEEMA), Brazil; Jardin Botanico, Rio de Janeiro, Brazil.

REFERENCES

Beck, B.B.; Martins, A.F. Golden lion tamarin reintroduction: Annual report for the Golden Lion Tamarin Conservation Project, Washington, D.C., National Zoo, 1995.

Beck, B.B.; Kleiman, D.G.; Dietz, J.M.; Castro, I.; Carvalho, C.; Martins, A.; Rettberg-Beck, B. Losses and reproduction in reintroduced golden lion tamarins (*Leontopithecus rosalia*). DODO: JOURNAL OF JERSEY WILDLIFE PRESERVATION TRUST, 27:50-61, 1991.

Beck, B.B.; Rapport, L.G.; Price, M.S.; Wilson, A. Reintroduction of captive-born animals. Pp. 265-285 in CREATIVE CONSERVATION: INTERACTIVE MANAGEMENT OF WILD AND CAPTIVE ANIMALS. P.J.S. Olney; G.M. Mace; A. Feistner, eds. London, Chapman and Hall, 1994.

Bronikowski, E.J., Jr.; Beck, B.B.; Power, M. Innovation, exhibition and conservation: free-ranging tamarins at the National Zoological Park. Pp. 540-546 in PROCEEDINGS OF THE AMERICAN ASSOCIATION OF ZOOLOGICAL PARKS AND AQUARIUMS ANNUAL CONFERENCE. Wheeling, West Virginia, American Association of Zoological Parks and Aquariums, 1989.

Golden Lion Tamarin Conservation Program. GLT Mission Statement. Washington, D.C., National Zoo, 1991.

Kleiman, D.G. Reintroduction of captive mammals for conservation. BIOSCIENCE 39:152-161, 1989.

Kleiman, D.G.; Beck, B.B.; Dietz, J.M.; Dietz, L.A. Costs of reintroduction and criteria for success: Accounting and accountability in the golden lion tamarin conservation program. Pp. 125-142 in BEYOND CAPTIVE BREEDING: REINTRODUCING ENDANGERED SPECIES TO THE WILD. J.H.W. Gipps, ed. Oxford, Oxford University Press, 1991.

Kleiman, D.G.; Beck, B.B.; Dietz, J.M.; Dietz, L.A.; Ballou, J.D.; Coimbra-Filho, A.C. Conservation program for the golden lion tamarins: Captive research and management, ecological studies, educational strategies and reintroduction. Pp. 959-979 in PRIMATES: THE ROAD TO SELF-SUSTAINING POPULATIONS. K. Benirschke, ed. New York, Spring-Verlag, 1986.

Miller, B.; Biggins, D.; Wemmer, C.; Powell, R.; Calvo, L.; Hanebury, L.; Wharton, T. Development of survival skills in captive-raised Siberian polecats (*Mustela eversmanni*), II: Predator avoidance. JOURNAL OF ETHOLOGY 8:89-94, 1990.

Miller, B.; Biggins, D.; Hanebury, C.; Vargas, A. Reintroduction of the black-footed ferret *(Mustela nigripes)*. Pp. 455-464 in CREATIVE CONSERVATION: INTERACTIVE MANAGEMENT OF WILD AND CAPTIVE ANIMALS. P.J.S. Olney; G.M. Mace; A.T.C. Feistner, eds. London, Chapman & Hall, 1994.

Vargas, A. Ontogeny of the endangered black-footed ferret (*Mustela nigripes*) and effects of rearing conditions on predatory behavior and post-release survival. Doctoral dissertation (UMI publication # 9430784), University of Wyoming, 1994.

Tara Stoinski is a PhD candidate at Georgia Tech University and a Research Associate at Zoo Atlanta and Collaborator on the Golden Lion Tamarin Conservation Program. **Dr. Benjamin Beck** is Associate Director for Biological Programs, National Zoological Park, Smithsonian Institute. He is a Member of the Management Group of Gorilla and Orangutan SSPs, and Great Ape TAG, Chair of the AZA Reintroduction Advisory Group, and Deputy Chair of the IUCN/SSC Reintroduction Specialist Group. Previously, Dr. Beck served as Curator of Primates and Research Curator at the Chicago Zoological Park (1970-1982). **Mary Bowman** is the GLT Project Coordinator at Zoo Atlanta and **John Lehnhardt** is the Assistant Curator of Mammals at the National Zoo.

A lion-tailed macaque (*Macaca silenus*) on display in the Trail of Vines exhibit at Woodland Park Zoo, Seattle. (Photo by Karen Anderson.)

STEADY-STATE PROPAGATION OF CAPTIVE LION-TAILED MACAQUES IN NORTH AMERICAN ZOOS: A CONSERVATION STRATEGY

Donald G. Lindburg, John Iaderosa, and Laurence Gledhill

Zoological Society of San Diego, San Diego, California (D.G.L.); St. Catherine's Wildlife Conservation Center, Midway, Georgia (J.I.); Woodland Park Zoo, Seattle, Washington (L.G.)

INTRODUCTION

From the first documented import of a lion-tailed macaque (*Macaca silenus*) in 1899 to the placement of restrictions on importation in 1970, zoological institutions in North America cooperated with one another primarily in exchanging individuals to meet local exhibition objectives. There was little interest in breeding, since replacements were readily obtained from wild stock [see Lindburg & Forney, 1992, for a summary of importations and captive breeding since 1899]. However, in response to reports of a declining wild population, zoos resolved nearly three decades ago to place greater emphasis on the captive breeding of this species [Hill, 1971]. By 1982, when a first international symposium on the global status of lion-tails was held, incentives were in hand to develop a management plan (Species Survival Plan©) for the North American population, under the aegis of the American Zoo and Aquarium Association [Foose & Conway, 1985; Gledhill, 1985].

During the 1980s, the North American population of lion-tailed macaques doubled in size [see Fig. 15.2 in Lindburg & Harvey, 1996] solely as the result of captive breeding (the last import of wild-caught animals was in 1968). An analysis of conception rates and infant mortality indicated that this rather dramatic increase arose from intensified effort and improved management by a relatively small number of zoos [Lindburg & Gledhill, 1992]. Additionally, computer simulations established that, given their life-history characteristics, a carefully managed population of 100 adults would be sufficient to conserve 90% of original gene pool heterozygosity over the next 100 years. This finding coincided with survey results indicating the availability of up to 200 spaces for the long-term maintenance of lion-tails in North American zoos. Having substantially exceeded that number in the years from 1983 onward, it became evident that population size should be reduced and a strategy developed for steady-state propagation, once the target size

Table 1. Genetic representation in the North American gene pool (N=194)*.

Mean Kinship	Designation	Number
0.000 - 0.0100	Very under-represented	11
0.0101 - 0.0250	Under-represented	48
0.0251 - 0.0399	Adequately represented	86
> 0.0400	Over-represented	49

* Does not include 29 individuals exempted from the management plan.

was attained. In this report, we describe the management plan that will guide future efforts in maintaining a stable population and indicate the rationale for the plan's components.

POPULATION ASSUMPTIONS
Genetic and Demographic Analyses

Current census figures indicate a lion-tailed macaque zoo population in North America of 223 individuals, of which 29 are reserved for special behavioral studies and are, accordingly, exempted from the management plan. The histories of all individuals were analyzed for the purpose of constructing pedigrees, identifying founders, and characterizing the genetic representation of each individual in the captive gene pool. The resulting mean kinship (MK) datum for each male and each female [Ballou & Lacy, 1995] provided a rank order for assigning breeding priorities. Approximately 30% were found to be under-represented, whereas about 25% have a mean kinship >0.0400, and can be considered as genetic surplus (Table 1).

Table 2. Management parameters for 1996 and subsequent years.

Available spaces in North American zoos	200
Annual recruitment rate (to reproductive age) needed to maintain steady state	10
Pregnancies/year required if reproductive rate is 50%	20

Table 3. Management categories for 1996.

Category	Number
Surplus	54
Designated to breed	19
Exempt (special research)	29
Hold for future breeding	121
Total	**223**

Captive lion-tailed macaques are non-seasonal in reproduction, and normally produce a single offspring from each pregnancy. The median birth interval for lactating females averages >17 months [Lindburg et al., 1989]. In order to maintain a core population of 100 adults with the level of founders currently in the population, a ratio of effective population size (Ne) to the actual population size (N) of 0.4 and a generation time of 12+ years has been established. By definition, effective population size is the number of individuals that would be required in a random breeding population of constant size and equal sex ratio to retain the same amount of genetic diversity as retained in the actual population [Ballou and Foose, 1996]. Experience indicates that only 50% of lion-tails survive to the time they are required for reproduction [Gledhill, 1990]. Calculations using these parameters indicate that an annual recruitment rate of 10 individuals, or 20 pregnancies, is required to maintain a stable population size (Table 2). However, it has been the policy of the management group to obtain fewer pregnancies in recent years, including 1996, until the desired population size is attained. Taking all factors together, the population is today subdivided into the four categories of surplus, current breeders, individuals held for future breeding, and individuals exempted from the management plan (Table 3).

Distribution

In addition to demographics, the distribution of individuals among the participating institutions was considered in developing a management plan. At the present time, 24 institutions hold lion-tails in numbers ranging from 3 to 30 (Figure 1). For the 1996 year, breeding was authorized in only 6 of these. The zoo population may be likened to groups having disjunct distribution in wild habitats, i.e., genetically isolated islands. Gene flow is, accordingly, realized by exchanges of individuals between collections (Figure 2). This process has two essential requirements: (1) use of genetic criteria in pairing to maintain the highest possible levels of genetic diversity in a population closed to the entry of new genes; and (2) simultaneously assuring social stability and reproductive viability in a popu-

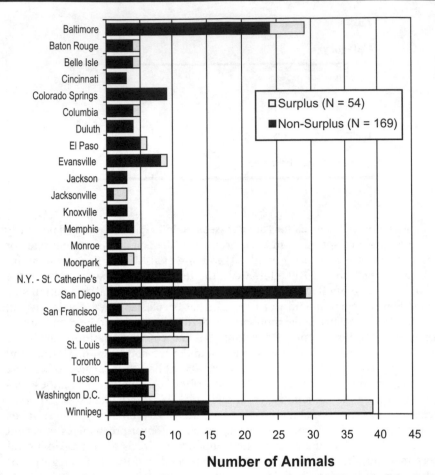

Figure 1. Distribution of lion-tailed macaques among institutions in North America (1996).

lation whose reproductive potential will not be fully exploited. Included under these requirements is the expectation that the full expression of mating behavior without incurring unwanted pregnancies will be realized, that aggressive potential will be held to acceptable levels, and that zoos may participate at several levels of involvement, depending on the resources they can allocate to this program.

BREEDING STRATEGY
Genetic Basis for Breeding Recommendations.

The highest priority is given to the breeding of individuals having the lowest mean kinship, i.e., those possessing the rarest alleles in the population. Mean kinship [Ballou & Lacy, 1995] was calculated by summing an individual's kin-

Figure 2. Gene flow in captive populations is achieved solely through animal exchanges. (Drawing by Sue Hohmann.)

ship coefficients with every other individual in the descent population, and dividing by the number of individuals in that population [see Lacy, 1993, for information on software]. Once calculated, a separate listing by gender from lowest to highest MK identifies the most desirable candidates for breeding from a genetic standpoint. An MK of ≤0.0200 was established as an initial cut-off point for making recommendations to pair individuals for mating.

Individuals with roughly similar MKs are preferred as partners in order to avoid combining rare with more common alleles in offspring, a procedure that would ultimately reduce genetic diversity. In addition, pairings that would produce inbred offspring are avoided. Other factors, such as age, physical condition, behavioral and reproductive history, and location are considered. For example, among those having the lowest MK values, priority is given to individuals whose alleles might soon be lost due to advanced age or poor physical condition.

Increasing the Intergenerational Time Span

The average intergenerational time span is at present placed at 14.4 years for males and 9.7 years for females, based on past breeding performance of the population [Gledhill, 1993]. Deliberate manipulation of breeders in order to increase generational length, e.g., by delaying the onset of reproduction, has the effect of slowing down the rate at which new individuals are recruited by birth into the breeding population. Through the end of 1982, studbook records indicate that infant mortality was significantly lower for multiparous than for first-time mothers [Gledhill, 1993]. According to Kumar and Kurup [1985], first reproduction by wild dams in India is estimated to occur at about 6 years of age. This contrast raises the possibility that the high-nutrition diets of captivity lead to early sexual maturation and subsequent dystocia in females not yet fully grown. Artificially delaying the onset of female reproduction to age 6 or later may, therefore, have the effect of increasing the generational time as well as the rate of infant survivorship.

A further recommendation of the breeding plan is that there be approximately two years between births for a given female, i.e., that an attempt be made to schedule births at ages 6, 8, 10, and 12, but allowing for some flexibility to age 15. Although there is no convincing evidence of a decline in lion-tail dams' fertility as they age, the potential for loss of a female through mortality obviously increases with age [Lindburg & Forney, 1992]. A female's potential for completing her genetic contribution to the population if reproduction is significantly postponed is, accordingly, diminished. Females whose infants die before the conclusion of the normal lactation/dependency interval of 12-15 months [Lindburg et al., 1989] will be at risk of conceiving earlier (since non-seasonal) than if estrus is suppressed by continued lactation. To keep such females on an approximate 2-year interval between births, they should be withheld from breeding until about 16 months after the previous parturition. Since initial post-lactation cycles are typically infertile [Clarke et al., 1993], a useful guideline in timing births is to allow for 5 1/2 months of gestation plus about 3 months, i.e., 3 cycles, for dams to become pregnant.

For captive sires, age of first reproduction has ranged from 3 to 21 years [Lindburg et al., 1989]. Because males have no direct involvement in infant care, there is seemingly no social consequence to the population if first reproduction by males is delayed to age 8. This age is closer to that realized by males in wild populations [Kumar & Kurup, 1985]. The rate at which individual males reproduce should be determined by their genetic value and by the females with whom they are to be mated. It is therefore theoretically possible for a male who begins breeding at 8 years of age to actually become reproductively surplus within a year or so. The more likely scenario is, however, that males will resemble their female counterparts in having reproduction spread out over a period of several years.

Social Considerations

Zoo management plans were originally driven primarily by genetic requirements [Foose et al., 1986]. In highly social species, it has become increasingly evident that group characteristics such as status hierarchies, social interdependency and stability, rearing environments, kin relationships, and aggressive potential must also be considered. The lion-tailed macaque plan reflects a major attempt at giving appropriate weight to these aspects of management in that it seeks to incorporate genetic objectives within a system that simultaneously promotes the social milieu so essential to primate well-being.

An initial premise in developing this plan is that no individual should be kept in social isolation except in the case of the strongly socially or physically disadvantaged or of those requiring isolation for medical treatment. The strategy which is described herein attaches paramount importance to the social needs of lion-tails.

A second premise is that females do not fare well when transferred between institutions, particularly when integration into an established social group is attempted. Not only do such individuals lose their former identity as members of a matriline [Birky, 1993], but they may undergo substantial trauma, possibly even death, during integration attempts. A decision has therefore been made to circulate only males among institutions for breeding purposes. To promote social stability and an appropriate socializing environment for offspring, males living permanently with females and their young in exhibit groups will be vasectomized to preclude unwanted breeding while at the same time allowing the full expression of sexual behavior to occur. Female candidates for breeding in a given year are to be temporarily separated from their group for short-term pairing during estrus with desired sires. Insofar as possible, traveling males will be accompanied by a female companion who is incapable of further reproduction. When fully operational, this plan will result in a permanent, stable core of females (and their offspring) in breeding/exhibit situations who will be reproductively serviced by breeding males that are rotated among institutions.

MANAGEMENT OF THE LOCAL UNIT (EXHIBIT AND/OR BREEDING GROUP)
Managing Females

Regardless of reproductive or genetic status, females are to remain in their natal groups until their demise unless there is a decision to establish a new group elsewhere or where females have been selected as companions for traveling males. Participants in this plan recognize that space problems can no longer be solved by moving small numbers of individuals of either gender elsewhere, as in former times. Reducing the size of exhibit groups will henceforth be accomplished by removing segments, e.g., matrilines, from a group to form a new group or by periodic removal of young males.

Managing Males

All males are to be removed from their natal groups before they become capable of reproducing, but not while they are still strongly dependent on their mothers, i.e., before their third but not before their first birthday. From a social standpoint, the tolerance of the alpha male for young males will vary with factors such as his disposition, the social standing of an immature male's dam, and enclosure size. The potential for some males to become reproductively active in advance of their third birthday must be carefully monitored.

Males that are surplus at birth, e.g., from accidental pregnancies, might be left in their natal group indefinitely, provided they are contracepted and remain compatible with other group members. Since successful establishment of all-male groups appears to be age-dependent, keeping sterilized surplus males in their natal group beyond the age of 3 or 4 will not be an option except in larger facilities and where excessive aggression does not become a problem. It should be recognized that late removal of males from the natal group usually leads to their long-term maintenance in social isolation or to euthanasia.

HOLDING MALES THAT ARE OUTSIDE THE BREEDING/EXHIBIT POPULATION

Since removal of males from their natal groups between the ages of one and three years is advocated, most males in the captive population will ultimately spend a portion of their lives outside the rotation system. In addition, it is clearly the case that genetic management of lion-tails results in individuals becoming reproductive surplus at various adult ages, i.e., as they meet their genetic quota [Lindburg & Lindburg, 1995]. Housing males outside breeding or exhibit situations will, with few exceptions, be required for both future breeders and reproductively surplus males.

Future Breeders

All males that are candidates for future breeding should be mother-reared to at least one year of age. Insofar as availability of facilities and local interests permit, a goal of the management group is to place as many of these males as possible into all-male groups of 2 to 5 individuals. Successful holding of males in one-sex groups is most feasible when groups are started while they are still socially immature, i.e., between the ages of one and three years of age. In institutions having relatively large social groups, it is conceivable that a point-in-time removal of 3 or 4 immature males from those group can be undertaken. More likely, however, males in this age group will be drawn from several institutions and placed in one that is prepared to undertake the assimilation process. Under this strategy, assuming such groups remain socially stable, individuals will eventually be removed as needed to join the rotation system for breeding.

Surplus Males

This category includes an adult male population on hand at the moment, as well as males that will become surplus after they have completed desired reproduction. Dealing with these males is more problematic, since most have (or will have) reached an age when they are too aggressive for holding in all-male groups [Clarke & Lindburg, 1989; Clarke et al., 1995]. It is desirable to minimize the impact of surplus males on space needs for the breeding population, and they should as a rule be rendered incapable of further reproduction. The possibilities envisioned for these animals are as follows: (1) As former breeders in the rotation system, they may continue to live with their companion females until they die or are euthanized for humane reasons; (2) Upon the demise of males presently in exhibit groups, a small number will be needed as replacements, in which case vasectomy is required; (3) Inclusion in all-male groups, including in some cases future breeders, should be attempted as resources and local circumstances permit; (4) Selective euthanasia, at the option of the holding institution, is recognized as a necessary course of action in some cases.

At the moment, the foregoing procedures for dealing with surplus males do not constitute totally sufficient options. The management group recognizes the importance of equalizing genetic contributions from both genders, although the necessity of establishing a one-male unit in order to obtain relief from inter-male aggression poses a major logistic problem for housing the majority of the males in the population. Given the option of devoting scarce space to species deemed to benefit from captive breeding, most zoos will offer little or no space to the holding of unisexual groups. Other factors mitigating against the keeping of "nonessential" macaques is their reputation for destructiveness, and concerns about transmission of *Herpesvirus simiae* to staff [Lindburg, 1993].

Euthanasia as a general solution should be avoided, if possible. On the other hand, options such as attempting to curb male aggressive potential through castration or gonadal down-regulation have not yet been proven viable. There is an obvious need for imaginative thinking and for research on ways to maintain surplus males with minimum impact on the cooperating institutions.

THE IDENTIFICATION OF SURPLUS INDIVIDUALS

An individual is deemed to be surplus when it is no longer required for any purpose, including genetic, social, and germ plasm contributions, in the North American population. Generally, the genetic contribution of individuals is deemed to have been satisfied after production of four offspring, two of whom have themselves reproduced. However, it is recognized that adjustments to this rule are required in the case of over-represented individuals, for whom the number to be produced will be lower, and under-represented individuals, for whom it will be higher. Once a male is classified as surplus, he is to be sterilized to prevent future reproduction if he remains in the North American population. It is recognized, however, that proven breeders that are surplus in North America may be desired for programs in other parts of the world. Although, in the majority of cases, it may be anticipated that an individual will have reached an age of 15 years or more by the time it becomes surplus, the variables that will apply in each case make age by itself an unsatisfactory criterion for designating an individual as surplus. Individuals that are genetically valuable but not reproductively viable may yet have a social value to the population, particularly in the case of aged adult females.

ASSISTED REPRODUCTION

At the present time, there is no requirement for utilizing artificial insemination, *in vitro* fertilization, or embryo transfer to facilitate reproduction in lion-tails. However, the management group recognizes that the plan here outlined may be subject to modification if techniques for assisted reproduction become routinely successful and can be used to simplify any of the procedures here advocated.

PAIRING PROCEDURES

Most mammalian populations in zoos serve both exhibit and breeding functions. For exhibited groups of primates, the usual course is for a heterosexual group to be established and for an intact, continuously present male to service the females of the group as they enter estrus. The timing and frequency of reproduction is thus dependent upon the proclivities of the animals themselves. When control of reproduction is necessary, contracepting individual females is often the course followed. In contrast, the system embraced by the lion-tail management group places an emphasis on social manipulation to regulate matings of genetically selected females with intact, genetically preferred sires. This process entails short-term removal of designated females from the natal group for pairing in a separate location during their period of receptivity. The question of when and how to stage pairings for breeding candidates therefore assumes major importance. Although the guidelines which are described here are useful in a general sense, it is recognized that each case may have unique aspects, depending upon individual temperaments and propensities. The prospects for success are therefore enhanced if pairings are entrusted to individuals who are familiar with the idiosyncrasies of the animals involved.

Additional concerns arise from the fact that high genetic ratings often derive from either behavioral or physiological deficiencies that have limited the prior reproductive output of the individuals so rated. Rearing deficiencies or hyper-aggressiveness, for example, may have precluded or reduced reproduction, causing individuals from lines that are not adequately represented in the gene pool to ascend on the priority list. The fact that the members of any given pair also have no prior experience with one another when males are rotated may also require that particular effort be invested in minimizing aggression between them.

The Timing of Pairings

The timing of the introductions is a critical factor. It is desirable on the one hand to minimize the time a female is separated from her natal group while at the same time insuring exposure to a potential sire to the extent needed for mating to occur. The follicular part of the cycle (tumescent phase of sex skin swelling), during which the ovum is maturing, averages about 13 days in lion-tailed macaques, whereas the fertile portion, when conception can occur, marks the end of folliculogenesis, and lasts about 48 hours [Shideler et al., 1983; Lindburg &

Harvey, 1996]. Two external cues aid in identifying the female's cycle stage: proceptive calling and sexual skin swelling and subsequent detumescence. Females begin to call with a staccato vocalization in early folliculogenesis (on average, 3 to 4 days after the start of menstruation) and will normally continue calling on a daily basis to approximately the time of ovulation. This call can be reliably used in the majority of cases (primiparae appear to call less regularly than multiparae) to predict the occurrence of the periovulatory phase within a few days [Lindburg, 1990].

Swelling and detumescence of the sexual skin can also be used to guide managers in selecting the best time for pairings. Usually, swelling begins to occur soon after menstruation. However, there are rather large differences between females or between cycles in the same female. For example, the amount of swelling in older females may be so slight as to be difficult to follow, and females coming out of lactation will usually have several somewhat irregular episodes of swelling before they are again fertile [Clarke et al., 1993]. In addition, some females are irregular as a matter of course, either in the duration of swelling or in the degree of swelling achieved in a given cycle. Again, the rule is to be familiar with the individual, and this requires that swellings be charted prior to pairing, using a numerical scale to rate the skin from flat to maximum swelling and subsequent detumescence. At its maximum size, the sexual skin becomes tightly stretched over the perineum, and has a shiny, smooth appearance.

Urinary monitoring of lion-tail cycles has established that detumescence of the sexual skin begins about two days after the mid-cycle peak in estrogen, i.e., just after ovulation has taken place [Lindburg & Harvey, 1996]. With the onset of detumescence, the sexual skin becomes less turgid, shows a softer, somewhat wrinkled surface, and loses its bright pink color. Mating that occurs after the onset of detumescence will in most instances be infertile. Because of individual and cycle-to-cycle variation, it is impossible to predict precisely the limited period when ovulation will occur. However, data from the San Diego colony indicate that females averaged about 7 days in the maximally swollen phase [99 cycles, Lindburg & Harvey, 1996]. In addition, mating in lion-tails invariably begins to occur before maximum swelling is achieved. Generally, then, one can commence pairing at the start of the maximum swelling phase and continue on a daily basis until detumescence begins. Males are less likely to be aggressive to females when they are partially or fully swollen than when flat.

The best strategy is to pair daily during the period marked by proceptive calling and sex skin swelling, with expectations of seeing 1 or 2 copulations within the first hour or two. Males have been known to achieve as many as 4 ejaculations within the first hour of pairing [Lindburg et al., 1985], but there is no proven relationship between frequency of copulations in a day and occurrence of conceptions. Lion-tails are serial mounters, i.e., males will mount numerous times before ejaculating. An ejaculatory mount is easily recognized as one in which the male remains maximally inserted but momentarily ceases to thrust. He appears rigid, and will sometimes utter a squeak vocalization coincident with ejaculation.

The average number of mounts to ejaculation is 9.1 [Lindburg & Harvey, 1996], but if exposure is limited to an hour or so per day, a male can sometimes ejaculate on the first mount [Lindburg et al., 1985].

Managing Aggressive Potential

Assuming the individuals to be paired have little or no familiarity with one another, it is recommended that several steps be taken to reduce the potential for male-inflicted trauma. The female should be given a day or two in the mating arena alone, and the male brought to her. If the male is disadvantaged by lack of familiarity with the area, he may be less inclined to exhibit unwanted aggression. However, too much novelty may also inhibit males from responding to the overtures of females, at least initially. The area designated for pairing should be sufficiently spacious and sufficiently equipped with climbing and perching structures to allow a female to avoid being cornered and attacked. Appropriate paraphernalia should be on hand to control aggression if needed [see Lindburg & Robinson, 1986, for suggestions for easing the trauma of introductions]. It is also recommended that the pair remain separated during periods when the staff is unable to monitor the course of events.

If a female stays maximally swollen during a pairing phase for more than a week, it may be advisable to return her to her group and try again during a later cycle. On the other hand, a female whose swelling begins to decrease after one or two days of copulation should be returned as soon as possible after onset of detumescence. There is as yet very little evidence to guide managers in returning a lion-tail female to her group after a period of separation, but temperament of the resident male and the female's social rank or presence of kin are factors that will likely influence that event.

Although more intensive management will be required to obtain pregnancies under this system, the extra effort for any given female is relatively infrequent. In the long run, a net gain in collection management efficiency should be realized through elimination of unplanned and unwanted births and the avoidance of work load increases that are required with rapid collection growth and the maintenance of a large surplus population.

Monitoring the Pairings

Participants in this breeding plan are expected to observe pairs during the time they are together in order to determine whether or not copulation occurs. They should also maintain records on other aspects of pairings that will be useful in evaluating the plan as a management strategy. The number of removals from her group that are required for pregnancy to occur should be recorded for each female. One cannot be certain that a female will ovulate during a given pairing episode, or that conception will occur (also, some high priority "breeders" may in fact be infertile). Early termination because of pair incompatibility or other management considerations may also preclude success in any particular episode.

A necessary aspect of this operation is that staff routinely monitor the sexual skin of females that are targeted for breeding. Pairings should take place while females are maximally tumescent. In some cases, a female may fail to achieve maximum tumescence or she may appear to detumesce prematurely. This result may be no more than natural variation between females or between cycles, or could be caused by stress related to the pairing itself. In any case, the result is that the mating attempt is canceled. Daily rating of the sexual skin during cycles in which females are paired for mating will contribute information useful in evaluating the rotating male system.

Tracking aggression or the avoidance of social contact are additional kinds of information that should be routinely collected. The two instances in which aggression may occur when using this system are: (1) between members of the mating pair when introduced, and (2) when females are returned to their home group. Careful attention to the recommendations for pairing should minimize male aggression toward females. However, at least some of the individuals that have high priority for breeding have had limited experiences with females, or may be limited by learning deficits or by temperaments that result in incompatibility. All instances of aggression leading to wounding or to termination of pairings either because of the risk of injury or failure to perform adequately (e.g., mounting deficiencies) or bond sexually should be noted. The latter includes cases in which one partner is fearful and avoids all attempts by the other at social contact.

Given that lion-tails live in multi-male groups in the wild and therefore change partners during a single estrus, it seems unlikely that a female returned to her group bearing the scent of another male will be attacked, but managers should be alert to this possibility. The time a female is away from her natal group should be kept as short as possible. Because hard data on this point are lacking for lion-tails, it is recommended for the moment that females not be absent for more than 7 days. All institutions should record the number of days females are absent, the response to females on their return to the home group, and any relevant group or individual circumstances that may have affected the response seen (for example, her social rank in the group, temperament of the resident male, time since the group was established, whether born into the group or added as a juvenile or adult, etc.). Depending on local circumstances, some institutions may want to ease females back into the natal group through social manipulation, e.g., placed with her kin initially, or with the resident male temporarily removed. As previously stressed, it is important that managers know the characteristics of individuals and take the precautions required to minimize injury.

It is unlikely that resident males will attack infants born into the group that have been sired several months earlier by an outside male. We know of no mechanism by which males may be able to distinguish between their own and another's progeny, particularly where there has been a long-standing and fairly continuous relationship with the mother. However, infanticide does occur in this species, and it is essential that detailed records on the circumstances of all infanticidal episodes be kept, so that a determination of cause can be deduced.

STERILIZATION PROCEDURES

Males that are selected for long term housing with the exhibit/breeding core of females should be sterilized via vasectomy. The Species Coordinator must give approval in each case in order to insure that only males that are truly surplus for breeding are used in this way. Vasectomy has no known effects on male libido [Phoenix, 1973], on males' ability to continue copulating (important to youngsters from a learning standpoint), or on maintaining order in the group. It is a procedure that is easily accomplished, and has a lower probability than castration of leading to medical complications. The time to loss of potency following vasectomy depends to some extent on how sexually active males are after surgery. It is suggested that a minimum of 30 days should pass, in any case, before males are introduced to females that are at risk for pregnancy.

Castration as a means of sterilization for males that may have a future in heterosexual groups is at present discouraged because of unpredictable effects on the males' subsequent behavior [Phoenix et al., 1973]. However, the management group is open to the possibility of using castration to diminish the aggressive behavior of males that are to be placed in all-male groups.

Females that are selected as companions for the rotating breeder males will consist of individuals that are diagnosed as incapable of further reproduction or surplus females that have been sterilized via tubal ligation. Ligation of the Fallopian tubes prevents pregnancy but not the production of sex steroids. It may be expected, therefore, that these individuals will continue to undergo cyclical sexual swelling and engage in copulatory behavior.

FACILITY REQUIREMENTS

Propagation of lion-tailed macaques under this plan requires a diversity of facility types, some or all of which may be available at any one institution. The maximum level of participation is as a breeding facility. This level requires a structure, either on or off exhibit, which is capable of housing 3 to 5 females, their progeny, and a vasectomized adult male as troop leader. In addition, the institution must provide a separate, long-term holding facility for a rotating male breeder and, usually, his female partner(s), and an arena for the staging of matings.

Zoos desirous of having lion-tails in their collections for exhibit purposes but not for breeding may participate in the plan either by maintaining breeding males and their companions while they are outside the breeding rotation, or as a holding/exhibit facility for an all-male group. The latter requires a space capable of holding 2 to 5 males, with a moderate to high degree of complexity such that they are provided with visual and physical barriers.

SPECIAL STUDIES IN BEHAVIOR

As noted in Table 3, 29 individuals in the North American population are at present exempted from the management plan, and reserved for "special research."

In 1982, participants in the first Lion-tailed Macaque Symposium charged with New York Zoological Society (now the Wildlife Conservation Society) and the Zoological Society of San Diego with the responsibility for preparing two captive born groups for reintroduction to native habitat in India [Heltne, 1985]. Accordingly, in 1989, a group from the San Diego Zoo was transferred to a newly constructed 0.3-hectare off-exhibit corral at its Wild Animal Park (WAP). This group was initially established at the Centre d'Acclimatization de Monaco in the early 1960s, and has been shown to have the matrilineal structure that typifies other macaques [Birky, 1993]. It is being maintained out-doors on a year-round basis, and fed in ways intended to encourage recognition of and foraging for foods likely to be encountered in the wild. Various forms of local wildlife, including snakes, bobcats, coyotes, deer, ground squirrels, skunks, and raptors are regularly seen in the surrounding area. The group is screened annually for viruses and other diseases by the Park's veterinary staff [see Lindburg & Harvey, 1996, for further details].

At its Wildlife Survival Center (WSC), located on St. Catherine's Island near Savannah, Georgia, the Wildlife Conservation Society established a free-ranging, radio-collared group in 1990 [Doherty & Iaderosa, 1993]. This group has been provisioned daily but, except for observational studies by students, human contact has been minimized. It is notable that the group has learned to forage for naturally occurring foods, and to range daily in an area near the provisioning center. The WSC group has also become increasingly arboreal and adept at climbing since its initial release. It regularly spends the night in the forest, and has coped successfully with a variety of local fauna, including potential predators such as raptors and alligators.

As these two groups have grown in size and adapted to their new circumstances, both efforts appear to be meeting the goals of the original charge. However, since the 1982 symposium, several new factors have come to light which bear on the original objective. The estimates of lion-tails remaining in the wild have increased from <1,000 to perhaps as many as 4,000. Upon further reflection, return of captive reared individuals to wild habitat has, accordingly, come to be viewed by some as a non-essential option. Experiences with captive lion-tails in India have often been unfavorable, and have generated little support from Indian professionals for reintroductions. It has been suggested, for example, that begging behaviors acquired in captivity are irreversible, and that without knowledge of predators there would be little chance that reintroduced populations would survive [Sankhala, 1977; Karanth, 1992]. Home range familiarity has been emphasized as elemental to long term survival [Kurup & Kumar, 1993], requiring knowledge of the timing and location of fruiting trees in order to reliably obtain food. There is also concern for the transfer of infectious diseases to the wild population [Sankhala, 1977], and field workers have suggested that reintroductions should be opposed because they divert precious financial resources away from saving wild populations [Karanth, 1992].

While many of these objections to a reintroduction project for lion-tails can be overcome through careful planning, it is clearly the case that its justification will be more scientific than conservationist based [Lindburg, 1992]. Although repatriation of the St. Catherine's and San Diego groups lacks immediate urgency from a conservation standpoint, the potential for limited experimentation before a crisis point in species' survival is reached is an option that may have merit. In addition to the potential for banking important information, a limited reintroduction effort would contribute to awareness of the plight of the wild population, and could create additional incentives within India to conserve lion-tailed macaque habitat.

Certain areas of otherwise favorable habitat in India, e.g., in Karnataka State, are devoid of lion-tails primarily because of hunting [Karanth, 1992]. If either relocation or reintroduction is utilized to repopulate these areas, a significant pre-release educational/ethnobotanical program would be required. Post-release protection and monitoring would be essential. Experience gained in establishing the groups at the WAP and WSC could be generalized to a reintroduction project. In addition, the adaptability demonstrated by these groups to more naturalistic conditions to date adds to the prospects for success of a reintroduction program for this species.

CONCLUSIONS

Controlling the growth of captive populations of lion-tailed macaques to fit into available zoo spaces while simultaneously meeting genetic and social objectives requires major adjustments in management protocols. A stable core of females residing permanently in participating zoos is a departure from the former practice of transferring females between institutions. Females designated for breeding in a given year are then serviced by genetically selected sires who are rotated among institutions, emulating the natural migratory pattern of the species in the wild. Several levels of institutional participation in this management plan are possible, e.g., propagating institutions, holders of all-male groups, or exhibition of non-breeding surplus for educational purposes. Among the management issues arising from the program are: the need for diversified facilities, developing criteria for the timing of pairings, managing the aggressive potential of participants, reducing the number of surplus individuals, and providing for the psychological and social well being of adult males that cannot be accommodated in exhibit groups. Groups of lion-tails that are candidates for repatriation to India have been established at the San Diego Wild Animal Park and the St. Catherine's Wildlife Survival Center.

ACKNOWLEDGMENTS

The authors thank their colleagues on the lion-tail management group who provided ideas and input in drawing up this plan: Sandra Kempske, Ingrid Porton,

Dennis Meritt, Jr., and Alan Shoemaker. We also thank Kevin Willis for early guidance in formulating this plan, and Dan Wharton, Jim Doherty, George Amato, Helena Fitch-Snyder, and Robert Lesnau, who participated in discussions leading to the plan's development.

REFERENCES

Ballou, J.D.; Foose, T.J. Demographic and genetic management of captive populations. Pp. 263-283 in WILD MAMMALS IN CAPTIVITY. D. Kleiman; M.E. Allen; K.V. Thomas; S. Lumpkin, eds. Chicago, University of Chicago Press, 1996.

Ballou, J.D.; Lacy, R.C. Identifying genetically important individuals for management of genetic diversity in pedigreed populations. Pp. 76-111 in POPULATION MANAGEMENT FOR SURVIVAL AND RECOVERY. J.D. Ballou; M. Gilpin; M.J. Foose, eds. New York, Columbia University Press, 1995.

Birky, W. FEMALE-FEMALE SOCIAL RELATIONSHIPS IN A CAPTIVE GROUP OF LION-TAILED MACAQUES (*MACACA SILENUS*). Master of Science thesis, Department of Biology, California State University, Northridge, 1993.

Clarke, A.S.; Lindburg, D.G. Can macaques live in all-male groups? ZOONOOZ 62(6):16-17, 1989.

Clarke, A.S.; Czekala, N.M.; Lindburg, D.G. Behavioral and adrenocortical responses of male cynomolgus and lion-tailed macaques to social stimulation and group formation. PRIMATES 36:41-56, 1995.

Clarke, A.S.; Harvey, N.C.; Lindburg, D.G. Extended postpregnancy estrous cycles in female lion-tailed macaques. AMERICAN JOURNAL OF PRIMATOLOGY 31:275-285, 1993.

Doherty, J.D.; Iaderosa, J.F. Free-ranging lion-tailed macaques (*Macaca silenus*) at the St. Catherine's Island Wildlife Survival Center. Pp. 90-95 in PROCEEDINGS OF THE AMERICAN ASSOCIATION OF ZOOLOGICAL PARKS AND AQUARIUMS ANNUAL CONFERENCE. Wheeling, West Virginia, American Association of Zoological Parks and Aquariums, 1993.

Foose, T.J.; Conway, W.G. Models for population management of lion-tailed macaque resources in captivity (a working paper). Pp. 329-341 in THE LION-TAILED MACAQUE: STATUS AND CONSERVATION. P.G. Heltne, ed. New York, Alan R. Liss, 1985.

Foose, T.J.; Lande, R.; Flesness, N.R.; Rabb, G.; Read, B. Propagation plans. ZOO BIOLOGY 5:139-146, 1986.

Gledhill, L.G. Progress toward a master plan of population management for the lion-tailed macaque. Pp. 379-383 in THE LION-TAILED MACAQUE: STATUS AND CONSERVATION. P.G. Heltne, ed. New York, Alan R. Liss, 1985.

Gledhill, L.G. LION-TAILED MACAQUE INTERNATIONAL STUDBOOK. Seattle, Woodland Park Zoo, 1990.

Gledhill, L.G. LION-TAILED MACAQUE INTERNATIONAL STUDBOOK (revised). Seattle, Woodland Park Zoo, 1993.

Heltne, P.G., ed. THE LION-TAILED MACAQUE: STATUS AND CONSERVATION. New York, Alan R. Liss, 1985.

Hill, C.A. Zoos' help for a rare monkey. ORYX 11:35-38, 1971.

Karanth, U.K. Conservation prospects for lion-tailed macaques in Karnataka, India. ZOO BIOLOGY 11:33-41, 1992.

Kumar, A.; Kurup, G.U. Sexual behavior of the lion-tailed macaque, *Macaca silenus*. Pp. 109-130 in THE LION-TAILED MACAQUE: STATUS AND CONSERVATION. P.G. Heltne, ed. New York, Alan R. Liss, 1985.

Kurup, G.U.; Kumar, A. Time budget and activity patterns of the lion-tailed macaque (*Macaca silenus*). INTERNATIONAL JOURNAL OF PRIMATOLOGY 14:27-39, 1993.

Lacy, R.C. GENES: A COMPUTER PROGRAM FOR THE ANALYSIS OF PEDIGREES AND GENETIC MANAGEMENT OF POPULATIONS. Brookfield, IL, Chicago Zoological Society, 1993.

Lindburg, D.G. Proceptive calling by female lion-tailed macaques. ZOO BIOLOGY 9:437-446, 1990.

Lindburg, D.G. Are wildlife reintroductions worth the cost? ZOO BIOLOGY 11:1-2, 1992.

Lindburg, D.G. Macaques may face a bleak future in North American zoos. ZOO BIOLOGY 12:407-409, 1993.

Lindburg, D.G.; Forney, K.A. Long-term studies of captive lion-tailed macaques. PRIMATE REPORT 32:133-142, 1992.

Lindburg, D.G.; Gledhill, L. Captive breeding and conservation of lion-tailed macaques. ENDANGERED SPECIES UPDATE 10(1):1-4, 10, 1992.

Lindburg, D.G.; Harvey, N.C. Reproductive biology of captive lion-tailed macaques. Pp. 318-341 in EVOLUTION AND ECOLOGY OF MACAQUE SOCIETIES. J.E. Fa; D.G. Lindburg, eds. Cambridge, Cambridge University Press, 1996.

Lindburg, D.G.; Lindburg, L.L. Success breeds a quandary: To cull or not to cull. Pp. 195-208 in ETHICS ON THE ARK: ZOOS, ANIMAL WELFARE AND WILDLIFE CONSERVATION. B.G. Norton; M. Hutchins; E.F. Stevens; T.L Maple, eds. Washington, DC, Smithsonian Institution Press, 1995.

Lindburg, D.G.; Robinson, P.T. Animal introductions: Some suggestions for easing the trauma. ANIMAL KEEPER'S FORUM 13:8-11, 1986.

Lindburg, D.G.; Lyles, A.M.; Czekala, N.M. Status and reproductive potential of lion-tailed macaques in captivity. ZOO BIOLOGY SUPPLEMENT 1:5-16, 1989.

Lindburg, D.G.; Shideler, S.E.; Fitch, H. Sexual behavior in relation to time of ovulation in the lion-tailed macaque. Pp. 131-148 in THE LION-TAILED MACAQUE: STATUS AND CONSERVATION. P.G. Heltne, ed. New York, Alan R. Liss, 1985.

Phoenix, C.H. Sexual behavior in rhesus monkeys after vasectomy. SCIENCE 179:493-494, 1973.

Phoenix, C.H.; Slob, A.K.; Goy, R.W. Effects of castration and replacement therapy on sexual behavior of adult male rhesuses. JOURNAL OF COMPARATIVE AND PHYSIOLOGICAL PSYCHOLOGY 84:472-481, 1973.

Sankhala, K.S. Captive breeding, reintroduction and nature protection: The Indian experience. INTERNATIONAL ZOO YEARBOOK 17:98-101, 1977.

Shideler, S.E.; Czekala, N.M.; Kasman, L.H.; Lindburg, D.G.; Lasley, B.L. Monitoring ovulation and implantation in the lion-tailed macaque (*Macaca silenus*) through urinary estrone conjugate evaluations. BIOLOGY OF REPRODUCTION 29:905-911, 1993.

Dr. Donald Lindburg is the Head of the Behavior Division at San Diego Zoo. He is a founding member and former President of the American Society of Primatologists, Editor of Zoo Biology and Associate Editor of the American Journal of Primatology. **John Iaderosa** is Curator-in-Charge of St. Catherine's Wildlife Conservation Center, located off the Coast of Georgia. He has been in this position for several years. St. Catherine's is owned and managed by the New York Zoological Society. **Laurence Gledhill** has retired as senior keeper of the primate section at the Woodland Park Zoo in Seattle, but currently works for the Zoo conducting conservation projects. He has been the studbook keeper and AZA species coordinator for lion-tailed macaques since the early 1980s and is the Macaque Subgroup Coordinator of the Old World Monkey Taxon Advisory Group (TAG) and a member of the TAG Steering Committee.

A drill (*Mandrillus leucophaeus*), one of Africa's most endangered primates. (Photo by Bill Johnston.)

DRILLS *(MANDRILLUS LEUCOPHAEUS)*: RESEARCH AND CONSERVATION INITIATIVES, 1986 - 1996

Cathleen R. Cox

Los Angeles Zoo, California

INTRODUCTION

Although the drill (*Mandrillus leucophaeus*) was reported to be in captivity as early as 1807 [Hill, 1970], the species remains an unfamiliar primate to many. In contrast, the drill's sole congener, the mandrill *(M. sphinx)*, is known to have been held in captivity as early as the middle of the seventeenth century [Hill, 1970], is now found in zoological gardens throughout the world, and is a favorite of zoo visitors. Both drills and mandrills are large, colorful, and highly sexually dimorphic forest-dwelling monkeys* (Figure 1). The perineal skin and hair of adult males is characterized by bright red and blue coloration and the facial skin of the adult male mandrill shows similar pigmentation. In contrast, the face of the drill is covered by a striking solid black mask. Historically, zoos have been less likely to display drills, perhaps due to the less flamboyant facial coloration. At present there are only 24 drills [Cox, 1996] compared to 186 mandrills [LaRue, 1996] in North America.

The drill was classified as endangered by USFWS in 1976, included in Appendix I of CITES in 1977, and first classified as endangered by IUCN in 1978. The IUCN/SSC Primate Specialist Group reviewed *in situ* conservation needs of African primates in 1986 and found drills among the six species in greatest need of conservation action [Oates, 1986a]. When the IUCN/SSC Primate Specialist Group updated the *Action Plan for African Primate Conservation* in 1996, the drill was the single species given the highest priority for conservation activity [Oates, 1996]. Despite the call for conservation action, drill population size continues to decrease, their habitat is becoming increasingly fragmented, and the large monkeys are being hunted extensively with the increasing demand for bush meat in

---

* Drills and mandrills are commonly referred to as baboons, but recent genetic research shows the genus *Mandrillus* to be more closely related to mangabeys in the genus *Cercocebus* than to baboons in the genus *Papio* [Disotell, 1994; Disotell et al., 1992; Inagaki and Yamashita, 1994].

Figure 1. An adult female and adult male drill (*Mandrillus leucophaeus*) at the Los Angeles Zoo. Note the extreme sexual dimorphism; adult body weight for females averages 32-37 lbs, whereas males may reach 65-90 lbs. in adulthood. (Photo by Heidi Engelhardt.)

African logging camps and urban centers. Clearly the survival of the drill is threatened. The 1986 *African Primate Action Plan* served to focus the attention of zoo biologists on the status of drills in captivity as well as in the wild. This chapter reviews work, carried out primarily during the past 10 years, that bears on the management of drills in captivity with initial steps to assist in conservation of drills in the wild.

STATUS IN CAPTIVITY AT THE BEGINNING OF THE DECADE

Mandrills reproduced in captivity as early as 1876, drills as early as 1910 [Jones, 1986]. In recent years, many zoo managers have limited mandrill reproduction as the number of births threatened to exceed the number of spaces available for additional animals. Drills were also known to reproduce well in captivity, though less frequently as there were fewer held in zoos. During the ten year period ending on 31 December, 1980, there were 28 drills born in North America, including two at the Los Angeles Zoo (LAZ). However, despite the breeding success at LAZ in 1979 and 1980, there were no births at the Zoo during the subsequent five years. Thus, zoo staff were prompted to explore the possible causes of this disturbing trend. Sixty hours of behavioral observation found no sexual activity and very few social interactions among the LAZ drills [Cox, 1987]. At the same time, Philadelphia Zoological Garden also noted an absence of births in their colony. Again, a 600 hour study of the drills found little evidence of mating and few other social interactions [Hearn et al., 1988]. For both zoos, the data indicated that lack of reproduction was a direct result of sub-optimal interactions among the drills. An informal survey of the status of drills in North America was conducted in 1977; there were seven institutions that housed drills but no drills were reproducing and the most recent birth of a youngster that lived at least 30 days had occurred in 1982.

These findings prompted a review of the annual reports distributed by the International Species Inventory System (ISIS), which confirmed an overall decline in the number of drills housed in zoological institutions. Moreover, examination of all volumes of the *International Zoo Yearbook* revealed that few youngsters survived to 30 days of age [Cox, 1987]. By the latter half of the 1980's, only the drills housed at the Hannover Zoo and the Wilhelma Zoo in Stuttgart were consistently producing young.

DEVELOPMENT OF THE DRILL SPECIES SURVIVAL PLAN©

In 1980, The American Zoo and Aquarium Association developed the Species Survival Plans© — a program which enables North American zoological institutions to manage species populations, making achievement of self-sustaining populations a high priority. Today there are 76 SSPs managing a total of 125 species. Primary criteria for identifying the need to develop an SSP include the classification of a species as threatened or endangered and the presence of that species in North American zoological collections. Additional information on the development and operation of SSPs is provided by Hutchins and Wiese [1991] and Wiese and Hutchins [1994, 1996, this volume].

Clearly, the drill would benefit from cooperative breeding management; the species is endangered in the wild and has a very small captive population with marginal reproduction. This realization was the basis for establishing an SSP for the drill. The proposal to form the drill SSP was approved by the Wildlife Conservation and Management Committee of the American Zoological Parks and Aquarium Association in 1988.

In 1989, representatives from all North American institutions that owned or housed drills generated an SSP masterplan designed to improve the future of captive drills and increase the likelihood of reproduction [Cox, 1989]. The SSP invited the participation of Hannover Zoo's Michael Böer, International Studbook Keeper [Böer, 1987] and drill European Endangered Species Program (EEP) coordinator. Böer's participation in the planning session brought invaluable information concerning the status of drills housed on other continents as well as the perspective of one institution where successful breeding was taking place.

Studbooks provide the lineage, birth dates, birth location, transfers and death dates of all animals that have been housed in captivity. Such records are essential for demographic and genetic analyses. As in most masterplan meetings, some of the historical studbook records were found to be incomplete and no further information could be obtained. In such cases creation of a "management" version of the studbook in which reasonable assumptions are made and then substituted for missing data played a central role. For example, if an individual's parents are unknown but the individual entered the captive population before a captive birth has been documented, that individual is assumed to have been captured in the wild. Review of the North American studbook data indicated that, if the continent's drill population was to become self-sustaining, both the number of founders contributing to the population and the rate of reproduction must be increased. Furthermore, the representation of founders would need to be equalized. A review of demographic parameters showed that accelerating reproduction and equalizing founder representation would be very difficult, as the population was aging. A similar review of the European drill population indicated it was in somewhat better condition but nowhere close to become self-sustaining.

Tracing the lineage of all members of the current drill population proved difficult on both continents. The number of known founders contributing to the population was five in North America, six in Europe. Only two founders had produced descendants that were housed in both North American and European institutions. Given the limited number of founders on the two continents, the masterplan organizers agreed that drills housed in North American and European institutions should be managed collectively, thereby increasing the combined number of founders to nine.

Based on the urgent need to improve drill reproduction, a number of transfers between institutions occurred prior to the 1989 meeting. For example, in 1986, four North American zoos transferred their entire drill stock to other SSP institutions to increase the average group size at the receiving institutions. In addition, four young animals were sent to North America from European zoos to provide a potential increase in genetic diversity. At the 1989 meeting, two more transfers of adult males from Europe to North America were recommended.

In each subsequent year since 1989, the SSP has considered the transfer of drills between institutions. In each case, the SSP must make the best decision to house together potential breeders who can produce offspring that will increase the representation of under-represented founders but have low inbreeding coefficients. This is done by maintaining an SSP management studbook that incorporates the current international data set, and then using SPARKS software [International Species Information System (ISIS), 1996] to calculate mean kinship of each living individual, as well as the inbreeding coefficients of offspring that could result from potential pairings. The drill population remains small, so every potential breeder is placed in a breeding situation and, to the extent possible, potential breeders housed together have similar mean kinships. Pairing individuals with equivalent mean kinships avoids the production of offspring whose subsequent reproduction would perpetuate unequal founder representation. Input from zoo representatives is essential to the SSP as familiarity with the behavior of individual drills and the perceived likelihood that a transfer would increase the propensity to breed is integral to making successful recommendations. In addition, moves are recommended only if the potential benefits of the transfer outweigh the potentially negative effects of travel and any quarantine period that would be required. A list of the December 31, 1996, mean kinships derived from the management studbook is shown in Table 1. Note that the mean kinship of each individual in the population depends on which population is being considered (North American, North American and European, or World) and changes with each birth; mean kinships must be recalculated prior to making a transfer recommendation.

The reorganization of drill group membership and subsequent transfers resulted in mating at four of the five North American zoos housing drills and, in 1992, the first successful North American births in a decade took place at LAZ and at Zoo Atlanta [Cox, 1992; Schaaf, 1992]. The Zoo Atlanta female that first gave birth in 1992 continues to reproduce, now having borne and reared a total of four offspring.

MONITORING BEHAVIOR

To assess the effects of management recommendations made by the SSP Committee, a protocol for collecting behavioral data was developed that is utilized by each North American institution that houses drills (available upon request from C. Cox, LAZ). In the protocol, a standardized sampling method is described. A staff member at each institution becomes an expert in drill data collection and trains others until they pass an interobserver reliability test. Observers watch drill groups for one-hour sample periods and record all bouts of certain behaviors that occur during the hour; at the same time, observers scan drill activity at one-minute intervals. Bouts of behavior are easily recognizable and can be reliably scored; included are 24 social behaviors, four scent-marking behaviors and five aberrant behaviors. As a basic operating practice, all institutions record these standard behaviors and some institutions score additional behaviors defined locally. All institutions use the same 15 scan categories with the option of subdividing categories into finer units if this meets the needs of a particular zoo. To facilitate learning the observation method, a 20-minute video tape, "The Behavior of the Drill: A Video Ethogram," was produced by the Greater Los Angeles Zoo Association [duBois, 1990]. The video tape is effective in familiarizing participants with drill behaviors and standardizing observations made at different zoos.

To facilitate data collection, several zoos use inexpensive but sturdy portable computers (Tandy 102s) equipped with bar code wands. An application written in BASIC specifically for drill data collection with these portable computers is used by each drill SSP institution. The program prompts the observer to record information according to the established protocol. Equipped with a portable computer and a sheet showing bar codes for initiators, recipients, behavioral bouts and activities, the observer records the standardized behaviors by entering the information directly into the computer. These files are then transferred to desktop computers, where the data are collated for statistical analysis.

appears that housing drills in one male, multiple female groups does not necessarily improve reproductive success.

Using the standardized protocol, observers compare the frequency of various behaviors as well as the percent of time spent in each activity. Researchers at the zoos in Los Angeles, Atlanta, Philadelphia, and Knoxville have all completed studies using this protocol. In addition, Erik Terdal used the same protocol in observation of the drills at zoos in Los Angeles, San Diego, Knoxville, Atlanta, Wuppertal, Hannover, Stuttgart, and Saarbrücken, and the mandrills in Portland, Chicago (Lincoln Park and Brookfield), Milwaukee, Madison, Lansing and San Francisco. The results of these studies are summarized below.

FINDING ADDITIONAL WAYS TO IMPROVE CAPTIVE REPRODUCTION

Group Composition

Hearn and her colleagues have focused on the comparison of reproductive success among drills housed in pairs as opposed to drills housed in groups of one adult male and two or more adult females [Hearn et al., 1991]. A review of historical records showed that, in the 1960's and 1970's, drills housed in pairs often produced young but many of these died as a result of parental neglect. In the 1980's, several zoos housed drills in one male, multiple female groups, thought to be more representative of drill social organization in the wild. With the exception of the drills housed in Hannover, however, the level of successful reproduction decreased in multiple female groups. In most other zoos, no reproduction occurred. At a few institutions, the top ranking female produced infants that survived, while subordinate females produced infants that died on the day of delivery. The deaths were thought to result from injury by other drills, but since births typically occur prior to dawn the actual trauma was never witnessed. Only at Hannover Zoo did housing multiple females with a single male result in highly successful reproduction; three to four females housed with a single male all produced young that were successfully reared. In summary, in the limited number of situations analyzed, it

1A.   MEAN KINSHIP & KINSHIP VALUES FOR THE POPULATION

Males		MK	MK	KV	KV	Age in	Proportion of
SB #/Name	Institution	Rank	Value	Rank	Value	Years	Lineage Known
142/*Michael*	Los Angeles	1	.0285	2	.0269	15	0.50
290/*Loon*	San Diego	2	.0392	1	.0264	17	1.00
314/*Bobby*	Atlanta	3	.0482	3	.0787	4	1.00
247/*Bart*	Knoxville	4	.0642	4	.0794	16	1.00
345/*Lyle*	Los Angeles	5	.0803	5	.1137	4	1.00
266/*Kurt*	San Diego	6	.0964	6	.1434	13	1.00
422/*Max*	Atlanta	7	.1160	7	.1752	2	1.00
295/*Adonis*	Atlanta	8	.1178	8	.1789	9	1.00
278/*Ace*	San Diego	9	—	9	—	21	0.00

Table 1. Data contained in the International Drill Studbook was obtained by linking ISIS records and information reported directly to the studbook keeper by zoos which hold or have held drills. At the time the studbook was first established, historical information on the parentage of a number of drills who had contributed to the current population could not be obtained and these individuals may or may not be founders. In such cases, the parents are entered as "unknown". When the number of known founders in a population is small, the genetic cost of not breeding animals of unknown parentage who are, in fact, founders is greater than the genetic cost of mistakenly assuming animals of unknown parentage are founders [Willis, 1993]. For this reason, management decisions are based on the Management Studbook which is a modified form of the International Studbook. In the Management Studbook all individuals with unknown parentage are carefully assessed and, in cases where it is likely that these individuals were brought in from the wild, the parents are entered as "wild". In the Management Studbook 14 of the 17 potential founders who are listed in the International Studbook as having unknown parentage are assumed to have been captured in the wild and are treated as founders. Only drills who have the physiological potential of reproducing are included in the tables.

Genetically important individuals can be effectively identified by calculating mean kinship (MK = average relationship of an individual to the population as a whole) and kinship value (KV = mean kinship weighted by the reproductive value of an individual's age class) from a management studbook [Ballou and Lacy, 1995]. In order to maintain expected heterozygosity as well as rare alleles in the population, reproduction of those individuals with low MK/KV should be encouraged and such individuals should be paired with others having similar values of MK/KV. MK can be obtained by utilizing GENES software [Lacy, 1993]; KV can be obtained by utilizing DEMOGRAPHY software [Bingaman and Ballou, 1996] prior to running GENES. MK and KV change whenever an individual is recruited into or removed from a population, whether by importation, birth, exportation, death or by initiating cooperative management between continents. Accordingly, MK and KV should be recalculated using the most current studbook data each time breeding recommendations are to be made. Below MK and KV are shown for the drills currently housed in North America as a function of different populations in which drills could be managed. In Table 1a the values shown were obtained by treating drills residing in North America as a separate population. In Table 1b the values shown are based on joint management of drills residing in North America and Europe. Table 1c shows current values of MK and KV that characterize the international population if all captive drills were to be managed jointly.

OF MALE AND FEMALE DRILLS HOUSED IN NORTH AMERICA

Females SB#/Name	Institution	MK Rank	MK Value	KV Rank	KV Value	Age in Years	Proportion of Lineage Known
281/*Melissa*	Los Angeles	1	.0357	1	.0037	22	1.00
286/*Becky*	Los Angeles	2	.0357	2	.0095	17	1.00
288/*Rosie*	San Diego	3	.0428	4	.0230	19	1.00
274/*Teal*	Knoxville	4	.0714	3	.0215	28	1.00
291/*Opal*	San Diego	5	.0785	5	.0273	17	1.00
292/*Pearl*	Atlanta	6	.0785	6	.0301	14	1.00
293/*Inge*	Atlanta	7	.0857	7	.1241	10	1.00
315/*Bioko*	Atlanta	8	.1160	10	.1751	4	1.00
440/*Ursula*	Atlanta	9	.1160	9	.1730	1	1.00
T9607/*Nora*	Atlanta	10	.1160	8	.1715	<1	1.00
320/*Ruby*	San Diego	11	—	11	—	24	0.00

1b.   MEAN KINSHIP & KINSHIP VALUES FOR THE POPULATION OF

Males SB#/Name	Institution	MK Rank	MK Value	KV Rank	KV Value	Age in Years	Proportion of Lineage Known
326/unk	Rabat*	1	.0000	1	.0000	7	1.00
327/unk	Rabat	2	.0000	2	.0000	5	1.00
301/*Congo*	Barcelona	3	.0125	5	.0169	11	1.00
142/*Michael*	Los Angeles	4	.0187	4	.0127	15	0.50
421/*Balombe*	Hannover	5	.0187	6	.0255	3	1.00
290/*Leon*	San Diego	6	.0203	3	.0096	17	1.00
345/*Lyle*	Los Angeles	7	.0578	8	.0619	4	1.00
247/*Bart*	Knoxville	8	.0585	7	.0611	16	1.00
313/*Gorbi*	Stuttgart	9	.0593	10	.0806	5	1.00
419/*Boris*	Hannover	10	.0593	9	.0804	3	1.00
259/*Viktor*	Hannover	11	.0750	11	.0863	14	1.00
266/*Kurt*	San Diego	12	.0812	12	.0934	13	1.00
310/*Adam*	Saarbrücken	13	.0835	14	.1037	6	1.00
422/*Max*	Atlanta	14	.0843	13	.0999	2	1.00
309/*Fritz*	Sofia	15	.0847	18	.1086	6	1.00
314/*Bobby*	Atlanta	16	.0847	19	.1087	4	1.00
318/*Florian*	Sofia	17	.0847	20	.1087	4	1.00
420/*Roman*	Hannover	18	.0847	17	.1085	3	1.00
T9612/unk	Hannover	19	.0871	15	.1039	<1	1.00
295/*Adonis*	Atlanta	20	.0906	16	.1061	9	1.00
426/*Jobst*	Hannover	21	.0925	21	.1092	1	1.00
442/unk	Saarbrücken	22	.0925	22	.1092	1	1.00
278/*Ace*	San Diego	23	—	23	—	21	0.00

* Although in Africa, the zoo in Rabat, Morocco, participates in the EEP.

MALE AND FEMALE DRILLS HOUSED IN NORTH AMERICA AND EUROPE.

Females SB#/Name	Institution	MK Rank	MK Value	KV Rank	KV Value	Age in Years	Proportion of Lineage Known
268/*Francoise*	Wuppertal	1	.0000	1	.0000	20	1.00
258/*Adelheit*	Saarbrücken	2	.0125	6	.0079	22	1.00
302/*Cabina*	Barcelona	3	.0125	11	.0169	10	1.00
281/*Melissa*	Los Angeles	4	.0156	2	.0013	22	1.00
286/*Becky*	Los Angeles	5	.0156	3	.0034	17	1.00
143/*Gail*	Saarbrücken	6	.0187	10	.0110	14	1.00
312/*Bubi*	Stuttgart	7	.0187	12	.0254	5	1.00
245/*LittleBit*	Saarbrücken	8	.0250	5	.0078	23	1.00
288/*Rosie*	San Diego	9	.0250	7	.0085	19	1.00
274/*Teal*	Knoxville	10	.0312	4	.0077	28	1.00
291/*Opal*	San Diego	11	.0343	8	.0098	17	1.00
292/*Pearl*	Atlanta	12	.0343	9	.0108	14	1.00
252/*Tschita*	Hannover	13	.0468	14	.0599	24	1.00
249/*Sonja*	Saarbrücken	14	.0515	13	.0599	22	1.00
254/*Heike*	Wuppertal	15	.0585	15	.0627	10	1.00
293/*Inge*	Atlanta	16	.0656	16	.0766	10	1.00
443/unk	Hannover	17	.0671	17	.0808	1	1.00
307/*Viktoria*	Wuppertal	18	.0835	21	.1019	7	1.00
315/*Bioko*	Atlanta	19	.0843	20	.0998	4	1.00
440/*Ursula*	Atlanta	20	.0843	19	.0991	1	1.00
T9607/*Nora*	Atlanta	21	.0843	18	.0985	<1	1.00
316/*Liza*	Hannover	22	.0867	25	.1067	4	1.00
433/*Daphne*	Hannover	23	.0914	23	.1040	1	1.00
T9609/*Rosi*	Hannover	24	.0914	22	.1035	<1	1.00
T9611/unk	Hannover	25	.0925	26	.1084	<1	1.00
255/*Sue*	Hannover	26	.0953	24	.1062	15	1.00
256/*Hanna*	Hannover	27	.0976	27	.1161	12	1.00
320/*Ruby*	San Diego	28	—	28	—	24	0.00

1c.   MEAN KINSHIP & KINSHIP VALUES FOR THE WORLD POPULATION

Males SB#/Name	Institution	MK Rank	MK Value	KV Rank	KV Value	Age in Years	Proportion of Lineage Known
332/*Jules*	DRBC	1	.0000	1	.0000	11	1.00
410/*Baba*	DRBC	2	.0000	2	.0000	10	1.00
338/*Ekki*	DRBC	3	.0000	3	.0000	9	1.00
411/*Bendeghe*	DRBC	4	.0000	4	.0000	8	1.00
326/unk	Rabat	5	.0000	5	.0000	7	1.00
413/*BillyOrif*	DRBC	6	.0000	6	.0000	7	1.00
342/*Rich.Iferi*	DRBC	7	.0000	7	.0000	7	1.00
340/*Kebi*	DRBC	8	.0000	8	.0000	6	1.00
398/*Petit*	DRBC	9	.0000	9	.0000	6	1.00
327/unk	Rabat	10	.0000	10	.0000	5	1.00
414/*Odey*	DRBC	11	.0000	11	.0000	5	1.00
343/*Osomba*	DRBC	12	.0000	12	.0000	5	1.00
416/*BasseyDuke*	DRBC	13	.0000	13	.0000	4	1.00
430/*Aniefiok*	DRBC	14	.0000	14	.0000	2	1.00
431/*Jono*	DRBC	15	.0000	15	.0000	1	1.00
T9603/*JoeForty*	DRBC	16	.0000	16	.0000	1	1.00
436/*Warrio*	DRBC	17	.0000	17	.0000	1	1.00
T9604/*Jegede*	DRBC	18	.0000	18	.0000	<1	1.00
301/*Congo*	Barcelona	19	.0081	22	.0106	11	1.00
331/*Base-Boje*	DRBC	20	.0081	21	.0088	7	1.00
444/*CreekTown*	DRBC	21	.0102	24	.0119	1	1.00
T9613/*PK*	DRBC	22	.0102	23	.0101	<1	1.00
142/*Michael*	Los Angeles	23	.0122	20	.0089	15	0.50
421/*Balombe*	Hannover	24	.0122	25	.0159	3	1.00
290/*Loon*	San Diego	25	.0133	19	.0067	17	1.00
432/*Bamu*	DRBC	26	.0286	29	.0337	1	1.00
T9605/*Sanje*	DRBC	27	.0286	28	.0333	<1	1.00
424/*Mgbochi*	DRBC	28	.0307	31	.0360	2	1.00
438/*Mbanjim*	DRBC	29	.0307	30	.0356	1	1.00
322/*Dezmondo*	Inuya	30	.0348	26	.0291	15	1.00
324/*Mylus*	Inuya	31	.0363	27	.0282	11	1.00
345/*Lyle*	Los Angeles	32	.0394	32	.0414	4	1.00
330/*BillyO'Ban*	DRBC	33	.0409	35	.0472	10	1.00
313/*Gorbi*	Stuttgart	34	.0409	39	.0533	5	1.00
419/*Boris*	Hannover	35	.0409	38	.0531	3	1.00
247/*Bart*	Knoxville	36	.0425	34	.0444	16	1.00
265/*Fredyl*	Osaka	37	.0450	36	.0508	11	1.00
439/unk	Osaka	38	.0453	33	.0434	1	1.00
T9610/unk	Wuppertal	39	.0468	37	.0510	<1	1.00
305/*Rudi*	Inuya	40	.0548	42	.0531	8	1.00
266/*Kurt*	San Diego	41	.0563	40	.0624	13	1.00
310/*Adam*	Saarbrücken	42	.0563	43	.0669	6	1.00
259/*Viktor*	Hannover	43	.0573	41	.0639	14	1.00
309/*Fritz*	Sofia	44	.0581	50	.0726	6	1.00
314/*Bobby*	Atlanta	45	.0581	48	.0725	4	1.00
318/*Florian*	Sofia	46	.0581	49	.0725	4	1.00
420/*Roman*	Hannover	47	.0581	47	.0713	3	1.00
422/*Max*	Atlanta	48	.0609	44	.0693	2	1.00
T9612/unk	Hannover	49	.0619	45	.0709	<1	1.00
295/*Adonis*	Atlanta	50	.0625	46	.0713	9	1.00
426/*Jobst*	Hannover	51	.0673	51	.0767	1	1.00
442/unk	Saarbrücken	52	.0673	52	.0767	1	1.00
278/*Ace*	San Diego	53	—	53	—	21	0.00

OF MALE AND FEMALE DRILLS HOUSED IN CAPTIVITY.

Females SB#/Name	Institution	MK Rank	MK Value	KV Rank	KV Value	Age in Years	Proportion of Lineage Known
268/*Francoise*	Wuppertal	1	.0000	1	.0000	20	1.00
415/*Comfort*	DRBC	2	.0000	2	.0000	4	1.00
417/*Tchika*	DRBC	3	.0000	3	.0000	3	1.00
T9602/*Pegi*	DRBC	4	.0000	4	.0000	3	1.00
423/*Noni*	DRBC	5	.0000	5	.0000	2	1.00
T9601/*Ekukunela*	DRBC	6	.0000	6	.0000	2	1.00
429/*Franca*	DRBC	7	.0000	7	.0000	2	1.00
412/*Kekere*	DRBC	8	.0040	11	.0047	8	1.00
329/*Calabar*	DRBC	9	.0040	14	.0048	8	1.00
335/*PepsiKola*	DRBC	10	.0040	10	.0043	7	1.00
334/*Alhaja*	DRBC	11	.0040	12	.0047	6	1.00
339/*Glory*	DRBC	12	.0040	13	.0047	5	1.00
258/*Adelheit*	Saarbrücken	13	.0081	17	.0055	22	1.00
302/*Cabina*	Barcelona	14	.0081	24	.0108	10	1.00
333/*ScarletteEbbe*	DRBC	15	.0081	22	.0091	6	1.00
337/*Elizabemi*	DRBC	16	.0081	23	.0091	5	1.00
281/*Melissa*	Los Angeles	17	.0102	8	.0009	22	1.00
286/*Becky*	Los Angeles	18	.0102	9	.0023	17	1.00
143/*Gail*	Saarbrücken	19	.0122	21	.0076	14	0.50
397/*Miki*	DRBC	20	.0122	25	.0144	8	1.00
312/*Bubi*	Stuttgart	21	.0122	26	.0159	5	1.00
245/*LittleBit*	Saarbrücken	22	.0163	16	.0050	23	1.00
288/*Rosie*	San Diego	23	.0163	18	.0056	19	1.00
274/*Teal*	Knoxville	24	.0207	15	.0048	28	1.00
291/*Opal*	San Diego	25	.0225	19	.0060	17	1.00
292/*Pearl*	Atlanta	26	.0225	20	.0069	14	1.00
437/*Afi*	DRBC	27	.0266	29	.0307	1	1.00
441/*Gaba*	DRBC	28	.0266	30	.0307	1	1.00
446/*Kikelomo*	DRBC	29	.0266	31	.0307	<1	1.00
435/*Sarebe*	DRBC	30	.0286	34	.0329	1	1.00
T9608/*Bambe*	DRBC	31	.0286	33	.0325	<1	1.00
T9606/*Matilda*	DRBC	32	.0307	35	.0351	<1	1.00
321/*JMC2*	Inuya	33	.0327	27	.0225	31	1.00
249/*Sonja*	Saarbrücken	34	.0338	36	.0379	22	1.00
252/*Tschita*	Hannover	35	.0348	37	.0435	24	1.00
311/*JMC7*	Inuya	36	.0376	32	.0316	5	1.00
241/*Ann*	Osaka	37	.0384	28	.0274	18	1.00
254/*Heike*	Wuppertal	38	.0445	38	.0474	10	1.00
443/unk	Hannover	39	.0502	40	.0582	1	1.00
293/*Inge*	Atlanta	40	.0512	39	.0569	10	1.00
307/*Viktoria*	Wuppertal	41	.0563	41	.0668	7	1.00
316/*Liza*	Hannover	42	.0584	45	.0697	4	1.00
315/*Bioko*	Atlanta	43	.0609	44	.0694	4	1.00
440/*Ursula*	Atlanta	44	.0609	43	.0689	1	1.00
T9607/*Nora*	Atlanta	45	.0599	42	.0684	<1	1.00
255/*Sue*	Hannover	46	.0645	46	.0713	15	1.00
433/*Daphne*	Hannover	47	.0645	48	.0720	1	1.00
T9609/*Rosi*	Hannover	48	.0645	47	.0721	<1	1.00
T9611/unk	Hannover	49	.0662	49	.0762	<1	1.00
256/*Hanna*	Hannover	50	.0680	50	.0803	12	1.00
320/*Ruby*	San Diego	51	—	51	—	24	0.00

The issue of drill social organization requires further attention. Because drills reside in forest habitat, they have proven quite difficult to study in the wild and the composition and structure of free-living groups remains unclear. Both Struhsaker and Gartlan published information on their sightings of drills in the late 1960's and early 1970's [Gartlan, 1970, 1975; Gartlan & Struhsaker, 1972; Struhsaker, 1969, 1972]. Gartlan [1970] illustrated the difficulty of observing wild drills; in the course of 15 months in the field, he had a total of 77.8 hours of "contact" with drills, but because the forest was dense, the primates were in "good view" for only 36.8 hours. Gartlan reported group sizes ranging from 14 animals, including one adult male, to 179 animals, including at least three adult males. He also provided evidence that the larger groups, called hordes, may form when several smaller groups, each associated with a single adult male, come together. Solitary males were also encountered.

Based on captive observations, the behavior and social organization of drills and mandrills appear to be very similar. As a result, information on mandrill social structure in the wild may be applicable to the captive management of drills. As described above, Gartlan reported occasional sightings of large groups of drills and Jouventin [1975] and Harrison [1988] reported even larger hordes of mandrills. Recently a horde of more than 600 mandrills was followed for several days [Rogers et al., 1996]. The age/sex composition of the horde was assessed three times; the mean ratio of the largest "fatted" males to smaller mandrills of both sexes was 1 to 21.2. The mean ratio of lean adult males and large subadult males to females and juveniles was 1 to 4.8. Repeated counts of smaller portions of the group confirmed that group composition was clearly multi-male.

Interviews with Nigerian hunters yielded numerous descriptions of multi-male groups of drills [Gadsby, 1990]. Based on these characterizations and the success of forming a multiple male semi-free-ranging captive group of mandrills in Gabon [Feistner et al., 1992], the majority of the young orphaned drills brought to the Drill Rehabilitation and Breeding Center (DRBC) in Nigeria were housed in a single group which proved to be very successful [Gadsby et al., 1994]. Initially, group size increased through the introduction of additional orphans and an open-air corral of approximately 200 m^2 was constructed to hold the group as they matured. The first birth at DRBC occurred in 1994 and, by the close of 1996, the group had 27 animals, including three adult males and seven adult females. All adult females are successfully reproducing. Based on this success, the zoos in Los Angeles and Atlanta are planning to form multi-male groups of drills.

The Importance of Parent-rearing

Prior to the establishment of the Drill SSP, North American zoos routinely separated expectant mothers from their group and provided human care for the infant if the mother showed neglect. In contrast, it has been the policy of European zoos to leave pregnant females with their mate or group and not to intervene in cases of maternal neglect. As a result of this policy, the majority of European drills have been reared in their social groups, and this may have led to substantial differences in parenting abilities. Terdal [1996] recently completed a detailed study of the possible factors associated with breeding success in captive drills and mandrills and he found that less active drills were less likely to engage in sexual behavior. Terdal predicted that human-reared drills would be less active than parent-reared drills and that parent-reared drills would be less active than wild-caught drills but his analysis did not support this prediction. Terdal believes that the lack of statistical significance may have resulted from the use of rearing history categories that were too broadly defined and did not consider the age at which animals were captured or the number of peers with whom they were reared. Forthman and Elder [1992] undertook a preliminary study to compare the behavior of four drills; one male and one female were nursery-reared and one male and one female were reared by the breeding female at Zoo Atlanta. The nursery-reared male was less likely to initiate social contact and the nursery-reared female was less likely to tolerate prolonged social contact. Cox, Hearn and Forthman are now collaborating on analysis of data collected at six zoos to further quantify the impact of nursery-rearing on drill aberrant behavior, social behavior, and reproductive potential. Variations in the style of nursery-rearing will also be assessed to identify ways to improve captive management.

Using Rewards to Increase the Likelihood of Social Interactions

The LAZ explored the use of positive reinforcement to increase the frequency of social interactions (including mating) among drills. The members of one group of drills were rewarded for exhibiting affiliative social behavior over an eight-month period. Rewards consisted of small pieces of their normal diet; a small hand-held clicker, clicked as food items were provided, became a conditioned positive reinforcer. Initially, rewards were given for allowing a subordinate drill to be fed undisturbed nearby. When this was achieved, the adult male drill was rewarded for gently touching the female with whom he had previously sired offspring [Desmond et al., 1987]. The training resulted in an increased frequency of social interactions during periods when the trainers were not present. However, actual mating was not achieved [Cox, 1988].

Improvements in Enclosure and Exhibit Design

A number of institutions have made modifications in their drill exhibits in an attempt to spark mating activity. Hearn et al. [1988] compared observations at the Philadelphia Zoo when the drills were restricted to a relatively small indoor area ($30 \, m^2$) to those made when the drills had access to a more spacious ($300 \, m^2$) and complex outdoor exhibit. The data showed an increase in the frequency of social behavior during the latter period. Such changes in habitat may favor reproductive success as they provide more 'things to do' and reduce the likelihood of intra-group aggression. In addition, a larger and more complex exhibit provides subordinate animals a greater opportunity to avoid aggression. However, the modification of the Philadelphia facility did not result in drill reproduction [Hearn et al., 1988]. Indeed, at the time the study was conducted in Philadelphia, Hannover Zoo had smaller (50 m²) and less elaborate enclosures than many North American institutions, yet maintained a much more consistent record of reproduction [Hearn et al., 1988].

In the wild, the forest habitat in which drills reside provides a plethora of visual barriers and these barriers may limit the frequency and/or intensity of aggressive interactions. Because such barriers may be a critical feature of drill habitat, researchers studied the effects of visual barriers on the behavior of two drill groups housed at Knoxville Zoological Gardens, each consisting of one male and two females. The first group had a history of little interaction and the low level of interaction was not influenced by the addition of barriers. There was a higher baseline level of social interaction among the second group which decreased with the addition of visual barriers [McMillan, 1991]. The presence of visual barriers may be particularly valuable in exhibits designed to hold larger groups or multi-male groups in which the level of agonistic interactions would be expected to be greater.

Environmental Enrichment

In 1992, the Philadelphia Zoo began an inexpensive keeper-initiated enrichment program to provide drills with a network of interlocking branches and a variety of large portable objects, such as telephone books and cardboard boxes. At the time, the zoo held three drills: a mature male, a cycling female, and a post-reproductive female. The male had a history of repeatedly injuring the cycling female and, as a result, these two were housed in separate enclosures and the post-reproductive female spent alternate days with each of the other drills throughout the study period. Provision of enrichment resulted in a substantial increase in social behavior as well as a significant decrease in aberrant behavior [Hearn et al., 1993].

Staff at the LAZ developed and tested six forms of environmental enrichment designed to increase the amount of time captive drills spend feeding and foraging, which is a major activity of wild drills. Enrichment consisted of food items scattered on the substrate or packed in simple but portable feeding devices. The scattered items included safflower seeds, meal worms, and crickets. The portable feeders included pine cones filled with peanut butter, boomer balls filled with sunflower seeds, and PVC tubing filled with crickets. Each form of enrichment was tested with two different groups of drills. Four of the forms of enrichment (scattered meal worms, scattered safflower seeds, sunflower seeds placed in boomer balls and crickets placed in cricket feeders) increased the amount of time spent feeding or foraging and reduced inactivity in the drill group. However, when enrichment was presented in the portable feeding devices, the frequency of agonistic interactions also increased. In both groups, the most dominant drill controlled the devices and spent the greatest amount of time feeding or foraging. The results of this study suggest that, when considering forms of enrichment for group-housed animals, it is important to consider the potential negative impact of competition within the group [Cox and duBois, 1992]. Overall, providing environmental enrichment improves the quality of life and, during the 12 month study, one of the LAZ drills conceived and delivered an infant that is now four years of age.

In his comprehensive multi-institutional study, Terdal [1996] recorded drill behavior and quantified a number of characteristics of the captive environment associated with 16 separate drill enclosures at eight different institutions. To assess environmental enrichment for drills, Terdal considered the extent of outdoor access, the presence of visual barriers, and the opportunity to climb on structures and manipulate objects. Environmental enrichment was positively correlated with the amount of time spent in active behaviors and, as described above, Terdal found a positive statistical relationship between activity level and sexual behavior.

Affiliative Interactions between Drills and Their Caretakers

Terdal also considered "keeper style" in his multi-institutional study and rated style on a 5-point ordinal scale, with the highest scores given to care-takers who hand-fed, groomed and/or trained the drills in their care with positive reinforcement. Affiliative husbandry style was positively associated with natural patterns of drill activity. Changing husbandry styles is a relatively simple means of improving drill management and the topic will be explored further by the drill SSP.

Exploring Similarities and Differences in the Behavior of Drills and Mandrills

Mandrills have been considerably more successful than drills in maintaining sustained reproduction in captivity. The reason for this success may simply be that more mandrills were imported from the wild and, by chance, the population adapted better to a captive environment. Captive mandrill groups rarely contain more than one adult male, but they tend to be held in larger groups than drills and, in that respect, they are managed in a manner more similar to their natural condition. Alternatively, there may be underlying species differences influencing breeding success. At this point, it is difficult to determine the reasons for the greater reproductive success of mandrills, but comparison of the two species and the conditions under which they are maintained promises to yield useful information. Terdal [1996] observed both drills and mandrills at 16 institutions. He found the impact of the captive environment on drill and mandrill behavior to be very similar and, as described above, he was able to determine the impact of a number of variables. In addition, Terdal's work provides additional evidence that the mandrill may serve as a suitable model for the more endangered drill.

Physiological Assessment of Reproductive Parameters

While we continue research on the social and environmental factors influencing reproduction, it is also important to investigate the potential for physiological limits on fertility. Zoo Atlanta conducted a study of semen and testis characteristics of drills, comparing them with the more reproductively successful mandrills. Among sexually mature males, sperm count and motility were significantly lower for drills. Lower sperm count does not necessarily result in infertility, but it is a factor that needs to be considered if assisted reproduction techniques are undertaken [Gould and Schaaf, 1994]. Gould and Schaaf are currently evaluating female hormonal assays obtained from urine collected daily from six drills and three mandrills. Preliminary results suggest that lack of reproduction in drills does not appear to result from physiological abnormalities (Gould and Schaaf, personal communication). In the sample population, there was no significant species difference in cycle length, patterns of hormone cyclicity, nor hormone levels – with the exception of PdG, which was significantly higher in drills than in mandrills. All sampled variables appear to fall within normal physiological limits.

Artificial Reproduction

To equalize founder representation of males that do not mate when housed with cycling females or cycling females that resist the advances of males, it may be necessary to use assisted reproduction techniques. The simplest of these methods is intravaginal artificial insemination, which is likely to require repeated attempts for success. With this understanding, LAZ keepers began training a cycling female drill to present for artificial insemination. Conditioning her sexually inactive male cagemate to provide semen on request has proven more difficult, but remains an option. Meanwhile, the San Diego Zoo has collected and frozen semen from the male drills housed there. Some of the frozen semen has been transported to LAZ, where two attempts at artificial insemination without anesthesia have taken place. Pregnancy has not resulted from these initial attempts, but the technique remains a potentially valuable one [Hogan, 1991].

DRILLS IN THE WILD

Since the development of the initial *Action Plan for African Primate Conservation*, several surveys have been conducted to verify the exact range of drill habitat. At present, the known range of drills extends from the Cross River in Nigeria to the Sanaga River in Cameroon, with a disjunct population on the island of Bioko, Equatorial Guinea. In addition, there have been at least two sightings of drills reported from Gabon [Blom et al., 1992]. If these sightings are substantiated, they will confirm yet another disjunct population, one that overlaps the range of mandrills. However, based on analysis of museum skins, Grubb [1973] presents a strong argument that drills and mandrills are allopatric.

Surveys. In 1986, the only known viable population of drills inhabiting the African mainland was thought to be confined to Korup National Reserve, Cameroon. Although drills occupied parts of Nigeria in the past, the species had not been seen for many years and was believed to have been extirpated. Likewise, drills had been reported on the island of Bioko, a former Spanish colony which subsequently became part of Equatorial Guinea; whether drills remained on Bioko was unclear.

A 10-week survey on the island of Bioko resulted in encouraging findings in 1986. Butynski and Koster [1994] confirmed that drills were present on Bioko; indeed, they were fairly common in the areas not frequented by hunters. Zoo Atlanta sponsored two subsequent expeditions to obtain more complete data and make recommendations for areas needing protection [Schaaf et al., 1990; 1992]. In the surveys of 1986 and 1990, the greatest number of drills were sighted in the Gran Caldera le Luba. However, when Hearn returned to Bioko in 1996, the frequency of drill sightings in the same area was down by nearly 50% [Hearn and Berghaier, 1996]. Similar findings resulted from a more extensive survey in January, 1997, suggesting a substantial decline in the number of drills inhabiting the Gran Caldera region (G. Hearn, personal communication). In addition to the drill, there are three sympatric primates on Bioko, Preuss's guenon *(Cercopithecus preussi),* Pennant's red colobus *(Procolobus badius pennanti)* and black colobus (*Colobus satanas*), all identified as subspecies or populations of particular concern in the revised *Action Plan for African Primate Conservation* [Oates, 1996]. Efforts to stem the loss of these species must be undertaken on Bioko island.

Survey of the Oban Hills region of Nigeria confirmed the presence of both drills and gorillas [Harcourt et al., 1988]. However, the population of each is relatively small. Gadsby [1990] subsequently interviewed hunters in the area and at the time estimated that fewer than 3650 drills remained in Nigeria. The same team is now assessing the current distribution of drills and the amount of suitable habitat remaining in Cameroon [Anonymous, 1992; E. Gadsby, personal communication]. Gadsby [1995] now believes that the entire range of the drill is less than 40,000 km^2 and that there may be as few as 3000 drills remaining in the wild [Gadsby and Jenkins, 1996].

CONSERVATION INITIATIVES

The need to establish conservation initiatives to protect drills in the wild has been well known for nearly a decade [Oates, 1986a, 1986b] and, in the revised *IUCN Action Plan for African Primate Conservation* [Oates, 1996], the drill is identified as the single species in greatest need of conservation action. Drills continue to be hunted and eaten in an unsustainable fashion [Gadsby, 1990; Gadsby and Jenkins, personal communication]. While some drills are consumed in local villages, the majority of the meat is taken to urban markets where it commands a higher price [Colell et al., 1994; Gadsby et al., 1994; Fa et al., 1995]. In addition, growing human settlements and concomitant cultivation are resulting in fragmentation of the remaining drill habitat. Such fragmentation reduces the chance of successful immigration and emigration between groups and increases the likelihood of species extinction [Gadsby et al., 1994].

While a number of areas have been set aside as reserves in Nigeria, Cameroon, and Bioko, Gadsby et al. [1994] cite insufficient governmental resources to enforce the protected status of the drill. In many cases, local residents are unaware of the government regulations. Projects that focus on increasing local awareness of the drill and the benefits of conserving wildlife, and increase the economic incentives for maintaining drill populations, have strong potential for halting the decline in wild drill populations.

The Drill Rehabilitation and Breeding Center

Throughout their range, hunters use dogs to tree groups of drills and then shoot as many animals as possible. In many cases, females fall to the ground with infants clinging to them and the infants are taken as pets. In 1991, the Drill Rehabilitation and Breeding Center (DRBC) was established in Calabar, Nigeria, in cooperation with the state Ministry of Agriculture and Cross River State Department of Parks & Wildlife. The goals of the DRBC are to provide information that promotes local interest in conservation, to rehabilitate former pet drills, and to initiate an *in situ* captive breeding program. The captive breeding program provides the potential to release drills in protected areas and to augment the *ex situ* captive population with potential founders, an action that would greatly bolster the likelihood of establishing a self-sustaining population on a continent other than Africa.

The DRBC operates in Calabar, an urban area located in Cross River State. A number of young drills were confiscated by wildlife officials and taken to the facility, others were donated by pet owners. As described in the previous section focusing on captive breeding, the DRBC has had phenomenal success in establishing a large social group of rescued drills. The first birth occurred in 1994, followed by eight births in 1995 and five in 1996. By the end of 1996, the DRBC held a total of 48 drills.

In addition to developing a successfully reproducing group of drills, the DRBC is making a substantial educational impact in the local community. Groups of school children and interested adults are given tours where they can view – for the first time – social groups of drills, once common in southeast Nigeria. Donations of equipment, labor, and funds are being sought successfully by the DRBC. A quarterly newsletter acknowledges these contributions and provides news of drill conservation developments in other locales. The DRBC also provides management advice to the Limbe Zoo in Cameroon, which holds eleven drills.

The DRBC is now completing an additional goal: establishing a permanent facility for semi-free ranging drills to be located in drill habitat. The Afi River Forest Reserve has been identified as an ideal location to release the drills. A four to six hectare enclosure will be constructed which is similar in design to the enclosure at CIRMF, Franceville, Gabon [Feistner et al., 1992]. The Afi Reserve is located 200 km north of Calabar and just 20 km from Cross River National Park (CRNP) where free-living drills are found. The facility for semi-free ranging drills will be in a locale where tourists can experience the wildlife in nearby CRNP, then visit DRBC to get a closer view of drills. Providing income to the local residents from whom the land is leased, as well as providing a source of employment, will substantially alter the economic motivation of local villagers to protect drills. If ecotourism increases in the nearby forested habitat, additional villagers may be employed as park rangers and guides [Anonymous, 1993; Gadsby, 1995].

The site for the permanent facility was leased from the three villages of Buanchor and a one-half hectare drill compound was constructed in 1996. The compound consists of forested land surrounded by an open-air corral; electrified wire prevents contact with the perimeter fencing, thus precluding potential escapes. In November, 1996, 27 drills were moved from Calabar to Buanchor (E. Gadsby, personal communication). Future plans include constructing a large drill compound, upgrading the road leading to Buanchor, and building a veterinary infirmary, education center, and housing for staff and visitors.

The DRBC also supports a community-based anti-poaching project in nearby Afi Massif (E. Gadsby, personal communication), which is particularly encouraging as there are a number of threatened or endangered species in the area. In addition to the drill, there are eight sympatric primates identified as needing conservation action in the revised *Action Plan for African Primate Conservation* [Oates, 1996]. These include: the angwantibo, (*Arctocebus calabarensis*), pallid needle-clawed galago *(Euoticus pallidus)*, Preuss's guenon *(Cercopithecus preussi)*, red-eared guenon *(Cercopithecus erythrotis)*, red colobus (*Procolobus badius*), chimpanzee (*Pan troglodytes*), and gorilla (*Gorilla gorilla*).

CURRENT SSP PRIORITIES

Despite the efforts described in this chapter, the increase in rate of captive drill reproduction during the past ten years remains minimal. At Zoo Atlanta, one pair of drills is reliably reproducing, but the seven remaining cycling adult females have not conceived in the past ten years. A major area of concern is the possibility that the marginal success in captive breeding has resulted from attempts to pair two different drill subspecies or crosses resulting from previous inter-subspecific pairings. Unfortunately, the geographical origins of the founders of the drill populations in North America and Europe were not recorded when they came into captivity. *M. l. leucophaeus* occurs on the African mainland and *M. l. poensis* occurs on the island of Bioko [Grubb, 1973]. These two subspecies are believed to have been separated for at least 10,000 years [Eisentraut, 1965]. Despite the separation, the two subspecies remain quite similar in appearance and cannot reliably be distinguished visually. Schaaf has obtained fur samples from museum specimens collected in known locations and from drills being sold at markets in Bioko, and the DRBC has contributed samples from drills residing in Cameroon and Nigeria. George Amato, of the Wildlife Conservation Society, is currently examining these samples to determine whether the two subspecies of drills can be identified through their mitochondrial DNA. Preliminary results suggest that the two subspecies can indeed be distinguished on this basis. In anticipation of conclusive results, fur and tissue samples have been collected from all drills residing in North America. Upon the completion of Amato's research, the SSP will determine the subspecific origin of each North American drill.

The success with multiple females producing young at the DRBC, where 27 drills are held in a single enclosure, suggests that optimal group size and composition needs to be reconsidered by the drill SSP. Would increasing the number of drills in a group and increasing enclosure complexity or size bring improvements in the reproductive activity of the second and third generation drills in North America? The creation of multi-male groups at LAZ and Zoo Atlanta promises to provide useful information for the future.

Given the low rate of reproduction and the inevitable aging of the current drill population, the SSP encourages continued research in assisted reproduction. The San Diego Zoo is developing a protocol for embryo transfer to be used with a relatively young diabetic female drill who cannot be expected to carry a pregnancy to term. To provide suitable care for the developing young without impinging on any other female drill's reproductive potential, the diabetic drill's embryos will be transferred to mandrill females that have previously been successful in producing and rearing youngsters. This work will provide a good test for the feasibility of applying assisted reproduction techniques to other drills.

The rationale for housing drills in zoological institutions includes creating an increased awareness of the difficulties drills are facing in the wild and strengthening public support for conservation initiatives. LAZ, Knoxville Zoological Garden and San Diego Zoo are all planning new exhibits that will incorporate educational graphics designed to increase visitor interest. The SSP encourages participating institutions to be responsive to the feedback provided by visitors and to continually incorporate innovative education and display concepts.

Member institutions of the Drill SSP have contributed funds and information in their areas of expertise to assist in the support of the DRBC's captive breeding program and *in situ* conservation efforts. Continuing to provide support for *in situ* conservation projects remains a very high priority for the SSP; further work on mainland Africa and the island of Bioko are encouraged.

CONCLUSIONS

In summary, the status of the drill remains precarious in the wild and in captivity, but there is still the potential to improve conditions. To ameliorate the situation, a wide range of projects are needed and these efforts should be integrated into a well-coordinated long-term international program. Conservationists seeking to stem the loss of drills met at the XVth Congress of International Primatological Society in 1996. The group agreed to develop an IUCN/SSC Primate Specialist Group Action Plan specifically for the drill which promises to substantially improve future prospects for the severely endangered drill.

ACKNOWLEDGMENTS

Drill conservation involves many dedicated individuals and institutions, and their continuing and creative work in diverse areas is invaluable. The energy and inspiration provided by William S. Bain and those cited in this chapter are much appreciated, as are the comments made by the anonymous reviewers. Development of the standardized method of collecting behavioral data and the computerized code that facilitates data collection, as well as production of the drill video ethogram, were made possible by an IMS Conservation Award #IC-00362-90 to the Greater Los Angeles Zoo Association.

REFERENCES

Anonymous. Drill survey in Cameroon. ORYX 26:177, 1992.

Anonymous. Drill update. ORYX 27:63-64, 1993.

Ballou, J.D.; Lacy, R.C. Identifying genetically important individuals for management of genetic variation in pedigreed populations. Pp. 76-111 in POPULATION MANAGEMENT FOR SURVIVAL AND RECOVERY, ANALYTICAL METHODS AND STRATEGIES IN SMALL POPULATION CONSERVATION. J.D. Ballou; M. Gilpin; T.J. Foose, eds. New York, Columbia University Press, 1995.

Bingaman, L.; Ballou, J. DEMOG 4.2:DEMOGRAPHIC MODELING PROGRAM. Washington, D.C., The National Zoological Park, 1996.

Blom, A.; Alers, M.P.T.; Feistner, A.T.C.; Barnes, R.F.W.; Barnes, K.L. Primates in Gabon - current status and distribution. ORYX 26: 223-234, 1992.

Böer, M. INTERNATIONAL STUDBOOK FOR THE DRILL (*Mandrillus leucophaeus*). Hannover, Hannover Zoo, 1987.

Butynski, T.M.; Koster, S.H. Distribution and conservation status of primates on Bioko Island, Equatorial Guinea. BIODIVERSITY AND CONSERVATION 3:893-909, 1994.

Colell, M.; Maté, C.; Fa, J.E. Hunting among Moka Bubis in Bioko: dynamics of faunal exploitation at the village level. BIODIVERSITY AND CONSERVATION 3:939-950, 1994.

Cox, C.R. Increasing the likelihood of reproduction among drills. Pp. 425-434 in PROCEEDINGS OF THE AMERICAN ASSOCIATION OF ZOOLOGICAL PARKS AND AQUARIUMS ANNUAL CONFERENCE. Wheeling, West Virginia, American Association of Zoological Parks and Aquariums, 1987.

Cox, C.R. Drills (*Mandrillus leucophaeus*): Social behavior and reproductive activity in captivity. INTERNATIONAL JOURNAL OF PRIMATOLOGY 8:552, 1988.

Cox, C.R. DRILL (*Mandrillus leucophaeus*) SSP MASTERPLAN. Unpublished report on file at the AAZPA Conservation and Science Office, Bethesda, Maryland, 1989.

Cox, C.R. Drill born at the Los Angeles Zoo. AAZPA COMMUNIQUE, September 15, 1992.

Cox, C.R. Drill (*Mandrillus leucophaeus*). Pp. 98-100 in AZA ANNUAL RE-
PORT ON CONSERVATION AND SCIENCE: 1995 - 1996. Bethesda, Maryland, American Zoo and Aquarium Association, 1996.

Cox, C.R.; duBois, T.T. Comparison of different types of behavioral enrichment
on the activity and social behavior of drill baboons. P. 557 in PROCEED-
INGS OF THE AMERICAN ASSOCIATION OF ZOOLOGICAL PARKS
AND AQUARIUMS ANNUAL CONFERENCE. Wheeling, West Virginia,
American Association of Zoological Parks and Aquariums, 1992.

Desmond, T.; Laule, G.; McNary, J. Training to enhance socialization and reproduction in drills. Pp. 435-441 in PROCEEDINGS OF THE AMERICAN
ASSOCIATION OF ZOOLOGICAL PARKS AND AQUARIA ANNUAL
CONFERENCE. Wheeling, West Virginia, American Association of Zoological Parks and Aquariums, 1987.

Disotell, T.R. Generic level relationships of the Papionini *(Cercopithecoidea)*.
AMERICAN JOURNAL OF PHYSICAL ANTHROPOLOGY 94: 47-57,
1994.

Disotell, T.R.; Honeycutt. R.L.; Ruvolo, M. Mitochondrial DNA phylogeny of the
old-world monkey tribe Papionini. MOLECULAR BIOLOGY AND EVO-
LUTION 9:1-13, 1992.

duBois, T.T. (producer) THE BEHAVIOR OF THE DRILL: A VIDEO
ETHOGRAM. Los Angeles, GLAZA/Vulture Videos, 1990.

Eisentraut, M. *Rassenbildung bei Säugetieren und Vögeln auf der Insel Fernando
Poo*. ZOOLOGISCHER ANZEIGER 174:37-53, 1965.

Fa, J.; Juste, J.; Perez del Val, J.; Castroviejo, J. Impact of market hunting on mammal species in Equatorial Guinea. CONSERVATION BIOLOGY 9:1107-1115,
1995.

Feistner, A.; Cooper, R.; Evans, S. The establishment and reproduction of a group
of semi-free ranging mandrills. ZOO BIOLOGY 11:385-395, 1992.

Forthman, D.; Elder, S. Activity in hand-reared and mother-reared pairs of captive
drills (*Papio leucophaeus*). AMERICAN JOURNAL OF PRIMATOLOGY
27:28, 1992.

Gadsby, E.L. THE STATUS AND DISTRIBUTION OF THE DRILL (*Mandrillus
leucophaeus)* IN NIGERIA. Unpublished report to the Nigerian Government,
Conservation International and WWF (Project #6738), 1990.

Gadsby, E.L. The FFPS drill project. FAUNA AND FLORA NEWS (3):1-2, 1995.

Gadsby, E.L.; Jenkins, P.D. Africa's endangered drill - Conservation and *in situ*
captive breeding. Abstract # 178 in ABSTRACTS, XVITH CONGRESS OF
THE INTERNATIONAL PRIMATOLOGICAL SOCIETY. Madison, Wisconsin Regional Primate Research Center, 1996.

Gadsby, E.L.; Jenkins, P.D.; Feistner, A.T.C. Coordinating conservation for the
drill (*Mandrillus leucophaeus*): endangered in forest and zoo. Pp. 439-454 in
CREATIVE CONSERVATION: INTERACTIVE MANAGEMENT OF WILD
AND CAPTIVE ANIMALS. P.J.S. Olney; G.M. Mace; A.T.C. Feistner, eds.
London, Chapman & Hall, 1994.

Gartlan, J.S. Preliminary notes on the ecology and behavior of the drill, *Mandrillus leucophaeus*, Ritgen, 1824. Pp. 445-480 in OLD WORLD MONKEYS: EVOLUTION SYSTEMATICS AND BEHAVIOR. J.R. Napier; P.H. Napier, eds. New York, Academic Press, 1970.

Gartlan, J.S. The African coastal rain forest and its primates - threatened resources. pp. 67-82 in PRIMATE UTILIZATION AND CONSERVATION. G. Bermant; D. Lindburg, eds. New York, John Wiley & Sons, 1975.

Gartlan, J.S.; Struhsaker, T.T. Polyspecific associations and niche separation of rain-forest anthropoids in Cameroon, West Africa. JOURNAL OF ZOOLOGY 168:221-266, 1972.

Gould, K.G.; Schaaf, C.D. Comparative semen parameters of the drill and mandrill. Pp. 169-174 in CURRENT PRIMATOLOGY, VOL 3: BEHAVIORAL NEUROSCIENCE, PHYSIOLOGY AND REPRODUCTION. J.R. Anderson; J.J. Roder; B. Thierry; N. Herrenschmidt, eds. Strasbourg, Universite Louis Pasteur, 1994.

Grubb, P. Distribution, divergence and speciation of the drill and mandrill. FOLIA PRIMATOLOGICA 20:161-177, 1973.

Harcourt, A.H.; Stewart, K.J.; Inahoro, I.H. NIGERIA'S GORILLAS: A SURVEY AND RECOMMENDATIONS. Lagos, Unpublished report to the Nigerian Conservation Foundation, 1988.

Harrison, M.J.S. The mandrill in Gabon's rain forest - ecology, distribution and status. ORYX 22:218-228, 1988.

Hearn, G.W.; Berghaier, R.W. CENSUS OF DIURNAL PRIMATE GROUPS IN THE GRAN CALDERA VOLCANIC DE LUBA, BIOKO ISLAND, EQUATORIAL GUINEA. Unpublished report to the government of the Republic of Equatorial Guinea, 1996.

Hearn, G.W.; Koch, C.M.; Weikel, E.C. Social behavior in a group of successfully reproducing drills (*Mandrillus leucophaeus*) at the Hannover Zoo. Pp. 629 in PROCEEDINGS OF THE AMERICAN ASSOCIATION OF ZOOLOGICAL PARKS AND AQUARIUMS ANNUAL CONFERENCE. Wheeling, West Virginia, American Association of Zoological Parks and Aquariums, 1991.

Hearn, G.W.; Onderdonk, D.; Rish, P. Effects of increased cage complexity on behavior in captive drills. Pp. 287-289 in PROCEEDINGS OF THE AMERICAN ASSOCIATION OF ZOOLOGICAL PARKS AND AQUARIUMS ANNUAL CONFERENCE. Wheeling, West Virginia, American Association of Zoological Parks and Aquariums, 1993.

Hearn, G.W.; Weikel, E.C.; Schaaf, C.D. A preliminary ethogram and study of social behavior in captive drills, *Mandrillus leucophaeus*. PRIMATE REPORT 19:11-17, 1988.

Hill, W.C.O. PRIMATES. COMPARATIVE ANATOMY AND TAXONOMY. VIII. CYNOPITHECINAE: PAPIO, MANDRILLUS, THEROPITHECUS. New York, Wiley-Interscience Publishers, Inc., 1970.

Hogan, M. Training for behavioral enrichment and species propagation. Pp. 629 in PROCEEDINGS OF THE AMERICAN ASSOCIATION OF ZOOLOGI-CAL PARKS AND AQUARIUMS ANNUAL CONFERENCE. Wheeling, West Virginia, American Association of Zoological Parks and Aquariums, 1991.

Hutchins, M.; Wiese, R.J. Beyond genetic and demographic management: The future of the Species Survival Plan and related AAZPA conservation efforts. ZOO BIOLOGY 10:285-292, 1991.

Inagaki, H.; Yamashita, T. An investigation of intergeneric relations among Papionini monkeys based on fine hair medulla structures. PRIMATES 35:499-503, 1994.

International Species Information System (ISIS). SPARKS 1.4. Apple Valley, Minnesota, International Species Information System, 1996.

Jones, M. Successes and failures of captive breeding. Pp. 251-260 in PRIMATES: THE ROAD TO SELF-SUSTAINING POPULATIONS. K. Benirschke, ed. New York, Springer-Verlag, 1986.

Jouventin, P. *Observations sur la socio-ecologie du Mandrill.* LA TERRE ET LA VIE 29:493-532, 1975.

Lacy, R.C. GENES: A COMPUTER PROGRAM FOR THE ANALYSIS OF PEDI-GREES AND GENETIC MANAGEMENT OF POPULATIONS. Brookfield, Illinois, Chicago Zoological Society, 1993.

LaRue, M.D. MANDRILL *(Mandrillus sphinx)* REGIONAL STUDBOOK. To-peka, Kansas, Topeka Zoological Park, 1996.

McMillan, G.C. The effect of environmental manipulation on drill baboon social behavior. P. 638 in PROCEEDINGS OF THE AMERICAN ASSOCIATION OF ZOOLOGICAL PARKS AND AQUARIUMS ANNUAL CONFERENCE. Wheeling, West Virginia, American Association of Zoological Parks and Aquariums, 1991.

Oates, J.F. ACTION PLAN FOR AFRICAN PRIMATE CONSERVATION:1986-90. Stony Brook, New York, IUCN/SSC Primate Specialist Group, 1986a.

Oates, J.F. African primate conservation, general needs and specific priorities. Pp. 21-29 in PRIMATES: THE ROAD TO SELF-SUSTAINING POPULATIONS. K. Benirschke, ed. New York, Springer-Verlag, 1986b.

Oates, J.F. AFRICAN PRIMATES. STATUS SURVEY AND CONSERVATION ACTION PLAN, REVISED EDITION. Gland, Switzerland, IUCN, 1996.

Rogers, M.E.; Abernethy, K.A.; Fontaine, B.; Wickings, E.J.; White, L.J.T.; Tutin, C.E.G. Ten days in the life of a mandrill horde in the Lopé Reserve, Gabon, AMERICAN JOURNAL OF PRIMATOLOGY 40:297-313, 1996.

Schaaf, C.D. Drill infant being mother raised at Zoo Atlanta. AAZPA COMMU-NIQUE, September 15, 1992.

Schaaf, C.D.; Butynski, T.; Hearn, G.W. THE DRILL *(Mandrillus leucophaeus)* AND OTHER PRIMATES IN THE GRAN CALDERA VOLCÁNICA DE LUBA: RESULTS OF A SURVEY CONDUCTED MARCH 7-22, 1990. Un-published report to the government of the Republic of Equatorial Guinea, 1990.

Schaaf, C.D.; Struhsaker, T.T.; Hearn, G.W. RECOMMENDATIONS FOR BIO-
LOGICAL CONSERVATION AREAS ON THE ISLAND OF BIOKO, EQUA-
TORIAL GUINEA. Unpublished report to the government of the Republic of
Equatorial Guinea, 1992.

Struhsaker, T. Correlates of ecology and social organization among African
Cercopithecines. FOLIA PRIMATOLOGICA 11:80-118, 1969.

Struhsaker, T. Rain forest conservation in Africa. PRIMATES 13:103-109, 1972.

Terdal, E. Captive environmental influences on behavior in zoo drills and mandrills (*Mandrillus*), a threatened genus of primate. Doctoral dissertation, Portland State University, 1996.

Wiese, R.J.; Hutchins, M. SPECIES SURVIVAL PLANS: STRATEGIES FOR
WILDLIFE CONSERVATION. Bethesda, American Zoo and Aquarium Association, 1994.

Wiese, R.J.; Hutchins, M. The evolution of the AZA Species Survival Plan. ANI-
MAL KEEPER'S FORUM 23:58-62, 1996.

Willis, K. Use of animals with unknown ancestries in scientifically managed breeding programs. ZOO BIOLOGY 12:161-172, 1993.

Dr. Cathleen Cox serves as the Director of Research at the Los Angeles Zoo and
chairs the Drill SSP. She contributes to the development of SSPs for a variety of
species as an active member of two of the AZA's scientific advisory groups, BHAG
(Behavioral and Husbandry Advisory Group) and SPMAG (Small Population
Management Advisory Group).

Javan gibbons (*Hylobates moloch*) at the Port Lympne Zoo Park, Lympne, United Kingdom. (Photo by Ernie Thetford.)

MULTI-DISCIPLINARY STRATEGIC PLANNING FOR GIBBON CONSERVATION IN THAILAND AND INDONESIA

Ronald Tilson, Katherine Castle, Jatna Supriatna, Kunkun Jaka Gurmaya, Warren Brockelman, and Schwann Tunhikorn

Minnesota Zoo, Minneapolis, Minnesota (R.T., K.C.), University of Indonesia (J.S.), University of Padjajaran, Indonesia (K.J.G.), Mahidol University, Thailand (W.B.), Thai Royal Forest Department (S.T.)

INTRODUCTION

The Gibbon Species Survival Plan© (SSP) is one of approximately 65 cooperative breeding and conservation programs administered by the American Zoo and Aquarium Association (AZA). Formed in 1990, the Gibbon SSP was preceded by several years of study, during which several parameters of captive and wild populations of gibbons were evaluated. In 1990, an advisory group of professional zoo biologists met to discuss these parameters, which included: (1) life history, demographic and genetic characteristics of eight species of gibbons managed in North American zoos, (2) IUCN/Species Survival Commission (SSC) Primate Specialist Group priority ratings for wild Asian primates [Eudey, 1987], and (3) the IUCN/SSC Conservation Breeding Specialist Group's (CBSG) priority ratings for captive Asian primates [Stevenson et al., 1992]. That 1990 meeting was followed by several workshops and conferences aimed at determining which gibbon species need most urgent conservation action and how this can best be accomplished. This chapter describes the findings of gibbon population surveys and the resulting working group recommendations, with special emphasis on gibbon species from Thailand and Indonesia.

BACKGROUND

Using the matrix of information listed above, the advisory group prioritized by species the captive management goals for gibbons in North America. It was determined that *Hylobates moloch, H. hoolock,* and *H. klossii* had a high priority for captive conservation, and if a breeding program was to be initiated, it had to meet several criteria for importation of founders as outlined below. Historically, there have been insufficient numbers of these species in North America for a SSP endorsed breeding program. *H. pileatus* had a high priority for conservation, but due to low numbers of animals in North America, a global captive breeding program was necessary. *H. concolor* had a high priority and it was recommended that the captive breeding program should be expanded due to sufficient number of founders. *H. agilis* and *H. muelleri* had a lower conservation priority and it was recommend that no breeding program be established for these species. *H. lar* and *H. syndactylus* have a lower priority for conservation, and because of sufficient numbers in North America, it was recommended that these species be managed for containment and not for expansion.

Other advisory group recommendations established minimum requirements for SSP endorsed captive breeding programs (CBP) and identified research needs. Proposed CBP typically involving importation of new founders would only be endorsed if: (1) new animals would not compete for resources with ongoing programs, (2) multiple institutions would be involved in the CBP, (3) sufficient numbers of animals were available to meet demographic and genetic needs of a small population, and (4) the country of origin was involved in the CBP. This recommendation was intended to curtail the "collection" of specimen animals and emphasize the need to manage these endangered species cooperatively. Two research needs were also identified: (1) the definition of gibbon subspecies using a genetic/molecular basis, song structure and pelage characteristics and (2) the identification of subspecific status of wild populations and individual gibbons in AZA facilities.

The Gibbon SSP placed a breeding moratorium on *H. lar* and *H. syndactylus* in 1991 until a genetic analysis into subspecific differences within species could be completed for the AZA populations. Funding was raised within the zoo community for the genetic analysis of these populations and the study was completed in early 1995.

With this groundwork, the Gibbon SSP extended beyond its primary regional responsibilities and recommended in 1991 that *in situ* gibbon conservation was a high priority for this program. They recommended that several potential projects in Thailand and Indonesia be reviewed for possible collaboration.

The family Hylobatidae is strongly represented in Thailand by three species: *Hylobates lar, H. pileatus* and *H. agilis*. A fourth species, *H. concolor*, is commonly found in captive collections because it is often smuggled into Thailand from its native range in nearby Vietnam. According to the IUCN's Primate Specialist Group [Eudey, 1987], all of these species are threatened and *H. pileatus* and *H. concolor* are critically threatened. No formal management plan for wild populations or for captive populations exists in Thailand for any of these species. This problem is further exacerbated by the continuing influx of captive gibbons. In Thailand, captive gibbons held as former pets, have been abandoned at Thai zoos which cannot accommodate them. Animal rights groups have used this situation, and the alleged sale of some gibbons to private dealers, for fund raising purposes but have not supported efforts to ameliorate the situation. Adverse international publicity sponsored by animal rights groups contributed to a climate in which the United States government placed trade sanctions on Thailand for noncompliance with CITES. Trade sanctions also affected the use of United States

Figure 1. Poster commemorating and advertising the conference for a Thai gibbon Population and Habitat Viability Analysis (PHVA), held in 1993. Pictured is a white-handed gibbon (*Hylobates lar*). (Photo by Warren Brockelman.)

federal funds for gibbon research and conservation in Thailand. At about the same time, USAID was withdrawn from Thailand and United States support for local conservation efforts disappeared.

To help resolve this dilemma, the Royal Thai Forest Department requested the AZA Gibbon SSP, and subsequently the IUCN/SSC CBSG, to conduct a Population and Habitat Viability Analysis (PHVA) to assess the risks of extinction in wild populations, to investigate the lack of a structured conservation program, and to resolve the growing crisis of too many captive gibbons in Thailand (Figure 1).

The family Hylobatidae is strongly represented in Indonesia by six species (and a number of subspecies): *Hylobates lar, H. syndactylus* and *H. agilis* in Sumatra; *H. muelleri* (and *H. agilis)* in Kalimantan; *H. klossii* in the Mentawai Islands off west Sumatra; and *H. moloch* in Java. Using IUCN criteria, all of these species are threatened, the Javan species in particular. No conservation strategy was in place with the Directorate General of Forest Protection and Nature Conservation (PHPA) and there were no organized captive breeding programs in the Indonesian Zoological Parks Association (PKBSI). Protected areas where these species live in Java are small and extremely fragmented, and the wild populations continue to be decimated for the pet trade and for food.

In the *IUCN/SSC Primate Specialist Group Action Plan for Asian Primate Conservation: 1987-1991*, it was recommended:

> Only fragmented pockets of moist evergreen forest remain in west (and central) Java, to which these two rare species are restricted. Human population pressure makes it impossible to declare new reserves. Upgrading of protection at Ujung Kulon/ Gunung Honje (768 km^2) and Gunung Halimun (400 km^2), where the largest populations of these endemics probably occur, are needed to conserve the two species. Population trends of both *Hylobates moloch* and *Presbytis comata* should be monitored, and the status of *P. comata fredericae* in central Java should be determined. The endemic subspecies *Trachypithecus auratus sondaicus* also would benefit from this action. [Eudey, 1987, p. 40]

It is arguable whether any or part of the above recommendation was ever implemented. Thus, the Indonesian Primatological Society (IPS) requested the IUCN/SSC CBSG to conduct a Population and Habitat Viability Analysis (PHVA) Workshop on Javan primates, specifically the Javan gibbon (*Hylobates moloch*) and the Javan langur (*Presbytis comata*). One of the goals of this workshop was to develop a structured conservation program for these two species, which could then be used as a model to develop similar programs for the entire primate community across all of Indonesia.

Range country scientists, conservation organizations, and wildlife authorities have struggled to develop collaborative conservation programs for wild gibbon populations. The primary goal of the PHVA in both Thailand and Indonesia was to develop a *Gibbon Action Plan* which would serve as a guide to protect remaining gibbon habitat, eliminate human-caused mortality, and maintain genetically viable, self-sustaining, free-ranging populations of gibbon species. In order to achieve this goal, it was necessary to understand the risk factors that affected the survival of these wild gibbon populations.

POPULATION AND HABITAT VIABILITY ANALYSIS (PHVA)

By definition, an endangered species is at risk of extinction. The goal in the recovery of such a species is to reduce this risk to some acceptable level, i.e., as close as possible to the background or "normal" extinction risk that all species face. To do this, it is important to improve our estimation of risk as a result of different management options, to improve objectivity in assessing risk, and to add quality control to the process through internal consistency checks [reviewed in Tilson & Seal, 1994].

In the last several years, the discipline of Conservation Biology has grown to fill the overlap of Wildlife Management and Population Biology. A set of principles loosely known as **"Population and Habitat Viability Analysis" (PHVA)** is powerful enough to improve the recognition of risk, rank relative risks, and evaluate options. It has the further benefit of changing part of the decision making process from unchallengeable internal intuition to explicit, and hence challengeable, quantitative rationales. One consequence of a PHVA is that critical aspects of the biology of a species can be identified, indicating where further knowledge may substantially increase our ability to predict the fate of a population, and where management actions to change population dynamics might be especially effective.

In widely distributed species, local populations may be lost, but are readily re-established from adjacent populations. A single or a few remnant populations isolated from any possible source of supplementation and recolonization typically will not survive indefinitely. Thus, it is not sufficient to protect a remnant population from those causes of decline that eliminated other populations. Rather, aggressive action must be taken to increase the population and to establish or re-establish additional populations. The goal is to extract a population out of the extinction vortex by returning its numbers, range, and diversity to such levels that normal population dynamics, including temporary local extinctions, preclude the extinction process.

A metapopulation is considered to be those animals, perhaps further divided into smaller populations, that exchange individuals sufficiently often so that each population exchanges, on average, at least one migrant per generation with other populations. Simulation modeling of some of the known causes of variability in gibbon populations was needed to estimate the size of wholly isolated populations needed for demographic stability. It is likely that populations containing fewer than 50 adults will not be stable over periods of 100-200 years. On the other hand, habitats unable to support 50 adults may be important to their recovery. Natural or managed migration can connect small remnant populations to those in other such habitat patches, thereby constituting a larger, stable metapopulation.

Because every isolated population is vulnerable to extinction (for example, a disease epidemic can decimate even a large population), species viability requires at least three populations sufficiently discrete to be subject to independent fluctuations in numbers. If each population has a moderately low probability of extinction, and if the populations fluctuate independently of one another, then it is highly unlikely that species extinction will occur.

The rationale for the demographic goals explained above applies equally to the genetic goals set forth here. Populations of approximately 50 unrelated breeding adults (and this number may be too small) maintain sufficient genetic variation to minimize inbreeding (expected mean increase in the inbreeding coefficient less than 1%) and, therefore, this number has been recommended as a minimum size for isolated populations of domestic livestock, and for short-term minimum population sizes of wildlife. If populations consist of fewer than 50 breeding adults, migration and immigration (one or more per generation) can genetically link these populations, possibly leading to a viable metapopulation.

Table 1. White-handed gibbon habitat areas in effective conservation units in Thailand.

No.	Units	Habitat	Total Area	Core Density (gr./km²)	Perimeter multiplier	Core area	No. of groups	Ne*
North:								
37	Lam Nam Pai WS	278	396	0.5	1	316	198	396
19	Nam Tok Mae Surin NP	118	113	0.5	1.5	49	25	50
44	Doi Chiang Dao WS	113	579	0.5	1	483	290	579
136	Salawin WS	579						
133	Om Koi WS	219						
35	Mae Tuen WS	209	428	1	1	345	345	690
22	Wiang Kosai NP	242	242	0.5	1	180	121	242
15	Phu Miang-Phu Thong	149	149	0.5	1	100	50	100
138	Lansang NP	109	109	0.5	1	67	34	67
67	Phu Luang WS	266	266	1	1	200	200	400
	Sri Satchanalai NP	227	227	0.5	1	167	83	167
	Total, North		2509			1907		
Northeast:								
5	Nam Nao NP	715						
137	Phu Khieo WS	1231	1946	1	1	1770	1770	3540
2	Phu Kradueng NP	189	189	1	1	134	134	268
1	Khao Yai NP	640	640	3	1.5	488	1465	2930
	Total, Northeast		2775			2392		
West-Southwest:								
63	Khlong Wang Chao NP	500						
44	Khlong Lan NP	261						
54	Mae Wong NP	219						
31	Umphang WS	1387						
120	Huai Kha Khaeng WS	575						
146	Thung Yai Naresuan WS	1056						
67	Khao Laem NP	1000						
59	Sri Nakharin NP	479	5477	2.5	2	4325	10812	21625
9	Sai Yok	530	530	1	1	438	438	876

Table 1. White-handed gibbon habitat areas in effective conservation units in Thailand (continued).

No.	Units	Habitat	Total Area	Core Density (gr./km^2)	Perimeter multiplier	Core area	No. of groups	Ne
1	Salak Phra	427	427	1	1	344	344	688
17	Mae Nam Phachi WS	512						
28	Kaeng Krachan NP	2624	3136	2.5	1	2752	6880	13760
	Total, West-Southwest		9570			7859		
South:								
30	Sadet Nykrom Maluang	274	274	1	1	208	208	416
2	Khlong Nakha WS	426						
69	Kaeng Krung NP	460						
56	Sri Phang Nga NP	255						
9	Khlong Saeng WS	814						
28	Khao Sok	476	2431	1	1.5	2135	2135	4300
35	Khlong Phraya WS	60						
30	Khao Panom Bencha NP	45	105	1	2	23	23	50
23	Khao Luang NP	407	407	1	1	326	326	650
42	Khao Pu-Khao Ya NP	479						
147	Khao Banthad WS	1058	1537	1	1	1380	1380	2760
	Ton Nga Chang WS	152	152	1	1	103	103	210
20	Thale Ban NP	200	200	1	1.5	115	115	230
	Tai Rom Yen NP	410	410	1	1	329	329	658
	Total, South		5516			4619		

*Ne = effective population size. All areas are in km^2.

Table 2. Pileated gibbon habitat areas in effective conservation units in Thailand.

No.	Units	Total Habitat	Core Density Area	Perimeter (gr./km²)	Core multiplier	No. of area	groups	Ne
	Southeast:							
8	Khao Khieo-Khao Chomphu WS	118	118	0.5	1	75	37	75
4	Khao Soi Dao WS	626	2	1	1	578	1156	2312
14	Khao Kitchakut NP	56	682	2	1	810	810	3900
15	Khao Ang Ru Nai	1000	1000	1	1.5	45	135	270
13	Khao Chamao NP	83	83	3	1			
	Total, Southeast	1883				1508	2138	
	Northeast:							
1	Khao Yai NP	1280	1280	2	1.5	1065	2130	4300
39	Thap Lan NP	1326	1735	1.5	1.5	1485	2230	4500
41	Phang Sida NP	409						
33	Huai Sala WS	341						
21	Khao Phanom							
13	Dongrak WS	234	575	1	1.5	431	430	860
13	Yot Dom WS	161						
53	Phu Chong Na Yoi	624	785	1	1.5	617	620	1200
	Total, Northeast	4375				3598	5420	
	Total, pileated	6258				5106	7560	15000

For long-term maintenance of the genetic variation necessary for adaptive evolution, minimum total population sizes of about 500 have been suggested for species protection. A population of this number of breeding adults will lose hereditary phenotypic variation no more rapidly than variation is regained by recurrent mutation (for gibbons this may mean a census population size of 2-3,000 animals per species). Thus, adequate genetic variation is always available for natural selection to adapt the population to novel stresses and environments.

PHVA Workshop objectives for Thailand and Javan gibbon populations included:
1. Estimate probable populations of gibbons in protected areas, the degree of fragmentation of these populations, and their probabilities for long-term survival with no intervention;
2. Determine population sizes required for various probabilities of survival and preservation of genetic diversity for specified periods of time;
3. Project the potential expansion or decline of gibbon population numbers due to environmental changes, habitat alteration and differing management plans;
4. Explore the role of exchanges among disjunct gibbon populations to maintain viable populations;
5. Identify field methods to monitor population status and assess habitat quality;
6. Evaluate the possible role of captive propagation as a component of the above management options; and
7. Produce a conservation strategy for gibbons that presents the results and recommendations of the PHVA.

THAILAND: GIBBON HABITAT AND POPULATION STATUS

Estimates of habitat and population numbers for wild *H. lar* and *H. pileatus* gibbons were derived by measuring the size of the forest, then estimating the extent and type of available habitat within each forest, and multiplying that figure by estimated gibbon densities for different habitats (established earlier in the workshop for lowland, hill, and submontane rain forest) [Brockelman et al., 1994]. These estimates resulted in a total population of approximately 110,000 *H. lar* living in 31 populations, and approximately 30,000 *H. pileatus* in eight separate populations. The information is presented in two comprehensive tables for *lar* and *pileatus* populations, that integrate total habitat available, assumed density of gibbon groups in core areas, and estimated gibbon populations for each conservation area (Tables 1 & 2) [Brockelman & Trisurat, 1994a].

Since large populations of Thai gibbons are highly vulnerable to poaching and are probably still declining, more surveys in core and peripheral areas need to be conducted, and regular monitoring of the numbers of breeding groups in selected areas need to be carried out at regular intervals. To do this, protected area personnel need to be trained in simple methods of censusing gibbons.

THAILAND: GIBBON LIFE HISTORY AND VORTEX ANALYSIS

The PHVA includes a set of recommendations for reduction of human-caused mortality, research, and management of the wild populations, as well as recommendations regarding the history of the population, the population biology and simulation modeling of the population. This modeling is based on the knowledge and assumptions of field biologists working with these species in the wild. A subset of these follows [Seal, 1994].

- The total *H. lar* population in protected areas in Thailand is currently estimated at about 110,000 individuals. Thirty-one separate populations have been identified and tabulated. Sixteen of these populations are estimated at 1000 or more individuals with estimated effective population sizes of 500 or larger. Five populations have 200 or fewer individuals. Eleven populations fall between 200 and 1000 individuals.

- Each of the 4 geographic regions of Thailand contain sufficient number of animals in protected area populations to be managed genetically and demographically as geographic units with no transfer of animals between the regions.

- The 16 populations with 1000 or more individuals are at essentially zero risk of extinction over 100 years if their habitat remains intact and if losses due to hunting are fewer than 5 female adults and 2.5 female young per 1000 population per year.

- The 11 populations of 200 to 1000 individuals are at essentially zero risk of extinction over the next 100 years if their habitat remains intact and if their losses due to hunting are less than 1 adult female and 1 infant female per 200 individuals per year. The loss of genetic heterozygosity in these populations will range from 0.3% to 0.6% over 100 years or 0.5% or less per generation. These populations *might* benefit from the addition (by translocation) of 1 or 2 suitably chosen individuals from neighboring wild populations every 20-40 years. They would not benefit genetically or demographically from the addition of individuals from the captive population.

- Populations of fewer than 50 animals have an extinction risk of up to 20% in 100 years, particularly if the species is subject to inbreeding depression. These populations may require genetic supplementation every 20-30 years. This could be accomplished by the addition of 2-3 individuals. If the population is below carrying capacity it has the potential to double in size in about 25 years by natural reproduction, depending upon local threats and chance events. If demographic extinction occurs, then the sites would be suitable for recolonization either by translocation or by reintroductions from a captive population.

- *Hylobates lar* in Thailand is eligible for reclassification from IUCN endangered to vulnerable or safe status if the protected areas and habitat are maintained and if the population levels remain the same or increase over the next 10 years. (Over the past 10 years, the rate of projected population decline solely from habitat loss was sufficient to classify the species as endangered in Thailand. The estimated additional removals from hunting would have reduced the growth potential of the total population and more seriously threatened local subpopulations.)

- The total *H. pileatus* population in protected areas of Thailand is currently estimated at about 30,000 individuals. Eight separate populations have been identified and tabulated. Six of these populations are estimated at 1000 or more individuals, with estimated effective population sizes of 500 or larger. One population has 200 or fewer individuals and one population has between 500 and 600 individuals.

- Each of the 2 regions (southeast and northeast) contain sufficient numbers of animals in protected area populations to be managed genetically and demographically as geographic units with no transfer of animals between the regions.

- *Hylobates pileatus* in Thailand may be eligible for reclassification from IUCN endangered to vulnerable status if the protected areas and habitat are maintained and if the population levels remain the same or increase over the next 10 years. (Over the past 10 years, the rate of projected population decline solely from habitat loss was sufficient to classify the species as endangered in Thailand. The estimated additional removals from hunting would have reduced the growth potential of the total population and seriously threatened local subpopulations.)

The **Genetic Aspects of Gibbon Management Working Group** considered issues of small populations, whether they are wild or in captivity, and how genetic management may help ensure their long-term survival. The development of the in-country capability for molecular genetic genotyping was recommended. Other recommendations concerned with the management of healthy wild populations stressed the value of knowing geographic providence of individual gibbons involved in reintroduction, translocation, or captive management programs [Woodruff & Tilson, 1994].

It was decided early in the workshop that the **Regional Captive Management Plan Working Group** should set guidelines for establishing a captive management program for Thai gibbons — regardless of their status in the wild — because they are native (one endemic) species, considered by the IUCN/SSC Primate Specialist Group as threatened, and because there are no organized captive management programs for these species in Thai zoos. Recommendations included establishing a Regional Captive Management Program in Thailand, beginning with *H. pileatus* and possibly expanding to *H. concolor*, and further training of Zoological Parks Organization (ZPO) staff in gibbon health, husbandry and Single Population Analysis Records Keeping System (SPARKS) studbook management [Nimmanheminda et al., 1994]. The working group on **Captive Management of Thai Gibbons** then developed a comprehensive set of gibbon management guide-

Table 3. Javan gibbon (*H. moloch*) habitat areas in effective conservation units in Java.

Units/Areas	Available Habitat (km2)	Status	Forest Type	# Groups Observed	# Individuals (3.3/group)
1. Ujung Kulon	30	NP	L/undisturbed	0	0
G. Payung*	85	NP	L/disturbed	11	36
2. G. Honje	235	NP	H/SM/undisturbed	16	53
3. G. Jayanti	<5	PF	L/disturbed	3	10
4. Lengkong	<5	PF	L/disturbed	4	13
5. G. Porang	<5	PF	L/disturbed	2	7
6. G. Salak	70	PF	H/SM/disturbed	7	23
7. Telagawarna*	<5	NR	H/SM/disturbed	0	0
8. Gde-Pangrango	50	NP	H/SM/undisturbed	28	90
9. G. Kancana	<5	PF	H/SM/disturbed	—	—
10. G. Malang	<5	NR	H/SM/disturbed	4	13
Takokak	—	—	—	—	—
Cadas Malane	—	—	—	—	—
11. G. Sanggabuana	50	PF	L/H/SM/disturbed	2	7
12. Bojongpicung	<5	PF	L/H/disturbed	1	3
13. Pasir Susuru	<5	PF	L/disturbed	2	7
14. G. Masigit	<5	PF	H/SM/disturbed	2	7
G. Falu	—	—	—	—	—
15. G. Simpang	140	NR/PF	H/SM/undisturbed	3	10
16. G. Tilu	30	NR	H/SM/undisturbed	3	10
17. G. Tangkuban Perahu*	—	—	—	0	0
18. G. Malabar*	—	—	—	0	0
19. G. Bukittunggul*	—	—	—	—	—
Gr. G. Pernhu (proposed)	—	—	—	—	—
20. G. Kendang	120	PF	H/SM/??	8	26
21. Gr. G. Limbung (proposed)	—	—	—	—	—
G. Wayang	85	PF	H/SM/undisturbed	3	10
22. G. Cikuray*	<5	NR	L/disturbed	0	0
23. Lwng Sancang	?	PF	H/SM/dist(prsd NR)	4	13
24. G. Slamet	12.8	NR	H/SM/disturbed	5	17
25. G. Masigit	27	GR	H/SM/M/undisturbed	2	7
26. G. Burangrang	—	NR	—	2	7
27. G.A. Takokak	—	NR	L	—	—
28. Cikepuh/Cibatang	—	GR	L	—	—
29. D. Janpong	—	PR	—	—	—
30. Cisolok	3	NR	L/H/disturbed	4	17 (actual)
TOTALS	**793 km²/18 areas (with gibbons)**			**Observed Groups: 116**	**Estimated Total: 386**

Table 4. Javan gibbon (*H. moloch*) habitat areas and population estimates in effective conservation units in Java.

No. Units	Total Size(km²)	Available Habitat	Occupied Habitat	Edge Effect	Core Area (km²)	Projected Pop. (ind)	Observed Est. Pop. (ind)
National Parks							
1. Ujung Kulon	—	—	—	—	—	—	0
G. Payung*	—	—	—	—	0	0	0
2. G. Honje	550	150(L)	50(L)	28	22	66	36
G. Halimun	400	296(w/edge)	—	28	296	908	75 (pooled)
military	—	—	—(H)	—	62<1000m	440	22 (Kunkun)
	—	—	—(SM)	—	234>1000m	468	53
3. Gde-Pangrango	150	50 (SM)	28	22	—	43	90 (pooled)
Production Forest							
4. G. Salak	—	70 (SM)	—	20	50	100	51 (pooled)
5. G. Sanggabuana	—	50	—	28	22	43	7
6. Gr.G.Perahu (proposed)	—	—	—	—	—	—	—
7. G. Kendang	250	—	120 (SM)	44	76 (SM)	152	29 (pooled)
Gr. G. Limbung (prop.)	—	—	—	—	75	338	6 (pooled)
now G. Wayang	100	85	—	—	37.5 (H)	263	—
	—	—	—	—	37.5 (SM)	75	—
Nature Reserve							
8. G. Simpang	150	110 (SM)	—	42	68	306	10 (pooled)
	—	—	—	—	34 (H)	238	—
	—	—	—	—	34 (SM)	68	—
TOTALS						**1956**	**304**

NP = national park; NR = nature reserve; PRF = production forest; GR = game reserve; * = no gibbons observed; L = lowland rain forest (1-3 gibbons/km²; H = hill rain forest (7 gibbons/km²); SM = submontane forest (2 gibbons/km²). Projected population estimates were derived by multiplying core area size by population densities typical for habitat type (see text).

Ujung Kulon National Park: The area available for gibbons in 1984 was estimated to be 85 km² with an edge effect included for eastern section. In 1992, it was estimated that 50 km² were occupied by gibbons. Kunkun surveyed the northern edge of this area in 1992 and found no gibbons. Two groups of gibbons have been identified in a western area and do not appear to be contiguous with the eastern populations.

Gede-Pangrango: Observed gibbon populations: Bedogol-Cibodas = 13; Situ Gunung = 5; Cimungkat = 9; Selabintana = 1. Total group = 28 groups x 3.2 individuals = 89.6 individuals.

lines to serve as the basis for implementing the ZPO recommendations [Arsaithamkul et al., 1994].

The working group on **Gibbon Diseases** acknowledged that infectious diseases pose a significant risk to gibbons, both for long-term maintenance of captive populations and for any suggested translocation or reintroduction program for wild populations. They provided general recommendations for disease control through quarantine procedures and testing, and identified several diseases that represented unacceptable risks to wild populations. A comprehensive document on gibbon health issues was presented as part of this workshop [Ratanakorn et al., 1994].

The **Gibbon Rehabilitation and Release Working Group** convened to discuss the mechanics of rehabilitation and release of gibbons. The issue of translocation was added as it has a definite impact on rehabilitation and release. Three separate subjects were considered: release site selection; selection of animals and rehabilitation management; and release and post-release follow-up. Release site selection was addressed by the **Selection of Gibbon Reintroduction Areas Working Group**. The group generated a set of specific criteria guiding selection of appropriate gibbons for participation in reintroduction programs, as well as a set of guidelines for rehabilitation, release and follow-up programs. Translocation of wild gibbons from one natural habitat to another was considered, but was deferred as a management option for future consideration [Brockelman & Trisurat, 1994b].

INDONESIA: GIBBON HABITAT AND POPULATION STATUS

The Indonesian PHVA focused primarily on the distribution, status and threats of wild populations of gibbons on the island of Java, and secondarily on the distribution and status of wild populations of Javan langurs [Supriatna et al., 1994]. Only gibbons are considered here. The workshop provided a unique opportunity to bring together Indonesian primate biologists who have censused, or are presently censusing, primates at sites in Java and elsewhere in Indonesia, Chiefs and PHPA staff of Ujung Kulon and Gunung Gede Pangrango National Parks, and international representatives from the IUCN/SSC Specialist Groups and Australian and North American zoos. The workshop concluded with the drafting of a *Javan Gibbon and Javan Langur Action Plan* [Supriatna & Tilson, 1994].

Estimates of habitat and population numbers for wild Javan gibbons were derived through consensus of the field biologists. There were two categories of estimates: one was derived from direct observation of all known individual family groups based on long-term presence in Halimun, Gunung Gede Pangrango, and Ujung Kulon National Parks in Java; the second was derived from estimating population numbers by measuring the size of the forest, then estimating the extent and type of available habitat within each forest, and multiplying that figure by estimated gibbon densities for different habitats (established earlier in the workshop for lowland, hill, and submontane rain forest). The first estimate (from direct observations) resulted in a total population of approximately 386 gibbons living in 21 sites; the second estimate (from available habitat extrapolations) resulted in a total population of about 1,957 gibbons in eight sites. In the second estimate, very small populations were discounted as not being viable and thus were not included. The first estimate resulted in numbers that were considered too conservative (because some gibbon groups were probably overlooked), while the second estimate resulted in numbers that were considered too large to be realistic (because this method implied complete habitat saturation, which was not substantiated by field observations of workshop participants). A consensus was reached that the best estimate of the real status of wild Javan gibbons lies somewhere between these two estimates, but nearer the conservative number (Tables 3 & 4) [Gurmaya & Supriatna, 1994].

INDONESIA: GIBBON LIFE HISTORY AND VORTEX ANALYSIS

The analysis for Indonesian gibbons relied primarily on data derived from the Thai Gibbon PHVA for the white-handed *(Hylobates lar)* and pileated *(H. pileatus)* gibbons, for which more comprehensive and long-term observations on population dynamics were available [Seal, 1994]. Because of the similarity of these species, life history characteristics for the Thai species were considered valid for the Indonesian species. Vortex modeling indicated that adult females are the most valuable members of a gibbon population and that the death of an adult female is the life history variable that has the greatest influence on increasing extinction rates for all population sizes. Because infants are captured by killing the mothers and because gibbons have a serially monogamous and family-oriented social system, the presence of infants in illegal trade clearly represent dead family groups

Some species of island birds and spring fish have survived demographic bottlenecks with fewer than 400 individuals. However, even if the approximately 400 gibbons left on Java were considered a single interbreeding population, their total numbers may not be sufficiently large enough to be considered an evolutionarily viable population. Because these gibbons live in multiple fragmented populations, they will need to be managed as a metapopulation through some form of genetic supplementation. Even populations of 200 or fewer individuals, in habitats that will not support a larger population, will require continued monitoring and periodic genetic supplementation. These small populations should be evaluated and monitored individually and suitable conservation management plans developed for their particular needs. The removal of one adult female with young per year from stable populations of 100 or fewer individuals (considered to be at or near maximum densities) will approximately double the risk of extinction. The protection against removal from small populations should be of the highest priority [Seal & Supriatna, 1994].

The seven populations of 10-26 gibbons are not viable over 100 years. The risk of extinction varies from 20 to 100% depending upon the amount of environmental variation. If these populations are to survive they will require active genetic and demographic management as part of the metapopulation. The 11 populations of fewer than 10 individuals are at high risk of extinction and need to be evaluated for more extreme management strategies, such as rapid habitat expansion, genetic supplementation, translocation, captive propagation, or a combination of these options [Seal & Supriatna, 1994].

The situation is not much improved even if we consider populations at the upper limit of almost 2,000 gibbons. Of the more than 30 sites in Java where gibbons previously existed, many sites no longer have gibbons. Their habitat now has been reduced so severely that there are only an estimated eight sites left in Java that are considered as effective conservation units for the species. Only Gunung Halimun National Park has the potential to maintain populations numbering 1,000 individuals. The next best available gibbon habitats, in Gunung Gede Pangrango and Ujung Kulon National Parks, Gunung Simpang Nature Reserve, and Gunung Wayang and Gunung Kendang Production Forests, have the potential to maintain gibbon populations in the 100's only. Thus, even if gibbons could increase their numbers to maximum densities in these remaining habitats, they would still need continued monitoring and genetic supplementation for long-term viability. These projections are only achievable with near-zero levels of poaching and further habitat degradation [Seal & Supriatna, 1994].

It was decided on the first day of the workshop that the **PKBSI Javan Gibbon Program Working Group** should set guidelines for establishing a captive management program for Javan gibbons regardless of its status in the wild because it is an endemic species of Java that is considered critically endangered by the IUCN/SSC Primate Specialist Group and because there are only 14 Javan gibbons in the PKBSI zoos. Recommendations included: establishing a Javan Gibbon Studbook and appointing a Studbook Keeper, developing a minimum captive population of 10 reproducing pairs of gibbons, preparing a gibbon husbandry manual, and training PKBSI staff in gibbon health and husbandry techniques [Ashari & Manansang, 1994]. A comprehensive set of gibbon management guidelines were also developed to serve as the basis for implementing the PKBSI recommendations [Manansang et al., 1994].

The working group on **Javan Gibbon Diseases** acknowledged that infectious diseases pose a significant risk to gibbons, both for long-term maintenance of captive populations and for any suggested translocation or reintroduction program for wild populations. They provided general recommendations for disease control through quarantine procedures and testing, and identified several diseases that represented unacceptable risks to wild populations [Sajuthi & Teare, 1994].

JAVAN GIBBON ACTION PLAN

There are several problems which challenge conservationists in the long-term management of the remaining habitat of wild Javan gibbon populations. Because of their restricted habitat, fragmented populations, and small population numbers, these highly endangered Javan primates will need wise conservation management strategies for their long-term survival. Problems identified during the workshop were:

- Inconsistent and incomplete database for censusing wild Javan gibbons;

- The need for improved training of PHPA and field staff in populations censusing techniques;
- Uncertain present and future status of current habitats with gibbon populations;
- Continued high levels of human encroachment and habitat degradation of protected areas;
- Insufficient law enforcement and lack of funds for habitat protection.

With these five problems in mind, the following prioritized recommendations were made to address immediate and critical conservation issues of Javan gibbon populations:

- Assess the current extent of gibbon habitat protected areas using all available technology – satellite imagery, current aerial photographs, available geological, vegetation and PHPA land-use forest status maps, geographic positioning system (GPS) units to ground-truth field observations – to develop a comprehensive geographic information system (GIS).
- Complete a Java-wide population and habitat survey for gibbons and langurs at all sites less than 5 km² in size, and at sites of uncertain land-use forest status. The Indonesian Primatological Society needs to form a research team comprised of individuals from universities, PHPA, NGOs and research institutions to assess sites that have been positively determined to have gibbon and langur populations. The 15 most significant Javan gibbon sites to be surveyed were identified and prioritized.
- Train and educate PHPA staff and local NGOs by IPS representatives and other primate professionals on how to census and monitor primate populations and how to collect ecological data on primates in their respective parks and conservation areas.
- Census gibbon populations annually and evaluate their habitat by trained PHPA staff and primatologists in the national parks and other conservation areas identified in the Java-wide census. This database will be linked with the National Biodiversity Network Database centered at the University of Indonesia.
- Increase public awareness and encourage local community participation in the conservation of gibbons and langurs by PHPA staff and NGOs. This recommendation is further expanded upon by recommendations of the Human Demography and Community Participation in Conservation Working Group.
- Integrate conservation management policies of PHPA to strengthen law enforcement in protected and important conservation areas identified as priority sites in the Java-wide census.

Given the low population estimates of wild Javan gibbons, their extreme fragmentation, their low reproductive potential, the continued encroachment and degradation of their habitat, and insufficient habitat protection and law enforcement measures, a captive management program under the direction of the PKBSI was recommended. It should be based on founders already in captivity, and not extracted from wild populations. Wild populations of less than 10 individuals (which are at high risk of extinction) should be evaluated for their possible role in translocation programs, other conservation programs and captive management programs [Supriatna & Tilson, 1994].

CONCLUSIONS

To assess the status, distribution, and threats to gibbon species living in Thailand and Indonesia, a multi-disciplinary strategic planning workshop was organized within both range countries. Using the set of principles loosely known as "population and habitat viability analysis," a structured conservation program was developed for *Hylobates lar* and *H. pileatus* in Thailand and *H. moloch* in Indonesia. The final product for each country was a *Gibbon Action Plan* which gives specific recommendations to conservation authorities on what to do next.

The total *H. lar* population in protected areas in Thailand was estimated to be about 110,000 individuals, making this species in Thailand eligible for reclassification from IUCN endangered to vulnerable or safe status – if the protected areas and habitat are maintained and if the population levels remain the same or increase over the next ten years. Similarly, the *H. pileatus* population was estimated at about 30,000 individuals, and thus they too may be eligible for reclassification like *H. lar*. The situation for *H. moloch* in Indonesia was much different. The total populations of *H. moloch* were estimated to be between 400 to 2,000 individuals, depending on the censusing technique. A consensus was reached that the best estimate of the real status of wild Javan gibbons lies somewhere between these two estimates, but nearer the conservative number. Thus, *H. moloch* should remain classified as critically endangered.

Given the low population estimates of wild Javan gibbons, their extreme fragmentation, their low reproductive potential, the continued encroachment and degradation of their habitat, and insufficient habitat protection and law enforcement measures, a captive management program was recommended. However, it was emphasized that the program be based on gibbons already in captivity and not be based on gibbons extracted from the wild. By contrast, gibbons in Thailand were considered less threatened, and only *H. pileatus* should be considered as a species for captive management. Because of its large numbers and much greater distribution over Thailand, *H. lar* was considered not in need of a captive management program. Controlled gibbon rehabilitation for captive *H. lar* was considered a more important conservation priority for the species.

ACKNOWLEDGMENTS

At the request of Khun Watana Kaeokamnerd, Deputy Director General of the Royal Forest Department, a Population and Habitat Viability Analysis (PHVA) Workshop for Thai gibbons was held at Khao Yai National Park on 26-29 April,

1994. The workshop was organized and conducted by Schwann Tunhikorn (Royal Forest Department), Warren Brockelman (Mahidol University), Usum Nimmanheminda (Zoological Park Organization), Ronald Tilson (Minnesota Zoo), and Ulysses Seal (IUCN/SSC Conservation Breeding Specialist Group, or CBSG). The Asia Foundation (Bangkok) sponsored all Thai NGO participants. Other supporting organizations include: the Calgary Zoological Society; Minnesota, Omaha and Milwaukee Zoos; and the European Endangered Species Program (EEP) – London, Twycross, Paignton, Fota WP, Edinburgh & Duisburg Zoos.

The Indonesian PHVA for Javan gibbons *Hylobates moloch* and Javan langurs *Presbytis comata*, held at Taman Safari Indonesia on 3-6 May, 1994, was organized by Jatna Supriatna (University of Indonesia), Kunkun Gurmaya (University of Padjajaran, Bandung), Jansen Manansang (Taman Safari Indonesia), Ronald Tilson (Minnesota Zoo), and Ulysses Seal (IUCN/SSC Captive Breeding Specialist Group—CBSG). Sponsors were: the Indonesian Primatology Society (IPS); the Indonesian Directorate of Forest Protection and Nature Conservation (PHPA); Conservation International (CI); Taman Safari Indonesia (TSI); Minnesota and Milwaukee Zoos and the AZA Gibbon Species Survival Plan© (SSP); Perth Zoo's Silvery Gibbon Project; and the European Endangered Species Program (London, Twycross, Paignton, Fota WP, Edinburgh and Duisburg Zoos). It was attended by over 50 participants, primarily members of the Indonesian Primatological Society, but also PHPA; the Biological Science Club; the Indonesian Zoological Parks Association (PKBSI); Pusat Studi Biodiversitas (UI); Yayasan Bina Sains Hayati Indonesia (YABSHI); the Primate Research Center at Bogor Agricultural University; the Captive Breeding and Primate Specialist Groups of the IUCN/SSC; the Perth (Australia), Milwaukee (USA), and Minnesota (USA) Zoos.

REFERENCES

Arsaithamkul, V.; Castle, K.; Gates, R.; Morris, D. Captive management of Thai gibbons. Pp. 53-60 in THAI GIBBON POPULATION AND HABITAT VIABILITY ANALYSIS. S. Tunhikorn; W. Brockelman; R. Tilson; U. Nimmanheminda; P. Ratanakorn; R. Cook; A. Teare; K. Castle; U.S. Seal, eds. Apple Valley, Minnesota, IUCN/SSC CBSG, 1994.

Ashari, D.; Manansang, J. Establishment of a PKBSI Javan gibbon program. Pp. 85-87 in JAVAN GIBBON AND JAVAN LANGUR POPULATION AND HABITAT VIABILITY ANALYSIS REPORT. J. Supriatna; R. Tilson; K.J. Gurmaya; J. Manansang; W. Wardojo; A. Sriyanto; A. Teare; K. Castle; U.S. Seal, eds. Apple Valley, Minnesota, IUCN/SSC CBSG, 1994.

Brockelman, W.; Ratanakorn, P.; Redford, T.; Morin, T. Gibbon rehabilitation and release. Pp. 87-90 in THAI GIBBON POPULATION AND HABITAT VIABILITY ANALYSIS. S. Tunhikorn; W. Brockelman; R. Tilson; U. Nimmanheminda; P. Ratanakorn; R. Cook; A. Teare; K. Castle; U.S. Seal, eds. Apple Valley, Minnesota, IUCN/SSC CBSG, 1994.

Brockelman, W.; Trisurat, Y. Thai gibbon habitat and population status. Pp. 17-22 in THAI GIBBON POPULATION AND HABITAT VIABILITY ANALYSIS. S. Tunhikorn; W. Brockelman; R. Tilson; U. Nimmanheminda; P. Ratanakorn; R. Cook; A. Teare; K. Castle; U.S. Seal, eds. Apple Valley, Minnesota, IUCN/ SSC CBSG, 1994a.

Brockelman, W.; Trisurat, Y. Selection of gibbon reintroduction areas. Pp. 83-86 in THAI GIBBON POPULATION AND HABITAT VIABILITY ANALYSIS. S. Tunhikorn; W. Brockelman; R. Tilson; U. Nimmanheminda; P. Ratanakorn; R. Cook; A. Teare; K. Castle; U.S. Seal, eds. Apple Valley, Minnesota, IUCN/ SSC CBSG, 1994b.

Eudey, A.A., ed. ACTION PLAN FOR ASIAN PRIMATE CONSERVATION: 1987-91. Riverside, California, IUCN, 1987.

Gurmaya, K.; Supriatna, J. Javan gibbon and Javan langur habitat and population status. Pp. 19-41 in JAVAN GIBBON AND JAVAN LANGUR POPULATION AND HABITAT VIABILITY ANALYSIS REPORT. J. Supriatna; R. Tilson; K.J. Gurmaya; J. Manansang; W. Wardojo; A. Sriyanto; A. Teare; K. Castle; U.S. Seal, eds. Apple Valley, Minnesota, IUCN/SSC CBSG, 1994.

Manansang, J.; Sumampau, T.; Teare, A.; Gates, R.; Castle, K. Captive management of Javan gibbons. Pp. 89-94 in JAVAN GIBBON AND JAVAN LAN-GUR POPULATION AND HABITAT VIABILITY ANALYSIS REPORT. J. Supriatna; R. Tilson; K.J. Gurmaya; J. Manansang; W. Wardojo; A. Sriyanto; A. Teare; K. Castle; U.S. Seal, eds. Apple Valley, Minnesota, IUCN/SSC CBSG, 1994.

Nimmanheminda, U.; Dumnui, S.; Methaphirat, S.; Kamolnorranath, S.; Arsaithamkul, V.; Castle, K.; Christie, S.; Gates, R.; Gates, D. Captive management plan for gibbons in Thailand. Pp. 51 in THAI GIBBON POPULA-TION AND HABITAT VIABILITY ANALYSIS. S. Tunhikorn; W. Brockelman; R. Tilson; U. Nimmanheminda; P. Ratanakorn; R. Cook; A. Teare; K. Castle; U.S. Seal, eds. Apple Valley, Minnesota, IUCN/SSC CBSG, 1994.

Ratanakorn, P.; Kamalnorranath, S.; Cook, R.; Teare, A. Gibbon diseases. Pp. 61-81 in THAI GIBBON POPULATION AND HABITAT VIABILITY ANALY-SIS. S. Tunhikorn; W. Brockelman; R. Tilson; U. Nimmanheminda; P. Ratanakorn; R. Cook; A. Teare; K. Castle; U.S. Seal, eds. Apple Valley, Minnesota, IUCN/SSC CBSG, 1994.

Sajuthi, D.; Teare, A. Javan gibbon diseases. Pp. 95-96 in JAVAN GIBBON AND JAVAN LANGUR POPULATION AND HABITAT VIABILITY ANALYSIS REPORT. J. Supriatna; R. Tilson; K.J. Gurmaya; J. Manansang; W. Wardojo; A. Sriyanto; A. Teare; K. Castle; U.S. Seal, eds. Apple Valley, Minnesota, IUCN/SSC CBSG, 1994.

Seal, U.S. Thai gibbon life history and Vortex analysis. Pp. 23-36 in THAI GIB-BON POPULATION AND HABITAT VIABILITY ANALYSIS. S. Tunhikorn; W. Brockelman; R. Tilson; U. Nimmanheminda; P. Ratanakorn; R. Cook; A.

Teare; K. Castle; U.S. Seal, eds. Apple Valley, Minnesota, IUCN/SSC CBSG, 1994.

Seal, U.S.; Supriatna, J. Javan gibbon life history and Vortex analysis. Pp. 43-75 in JAVAN GIBBON AND JAVAN LANGUR POPULATION AND HABITAT VIABILITY ANALYSIS REPORT. J. Supriatna; R. Tilson; K.J. Gurmaya; J. Manansang; W. Wardojo; A. Sriyanto; A. Teare; K. Castle; U.S. Seal, eds. Apple Valley, Minnesota, IUCN/SSC CBSG, 1994.

Stevenson, M.A.; Baker, A.; Seal, U.S., eds. GLOBAL CAPTIVE ACTION PLAN FOR PRIMATES. Apple Valley, Minnesota, IUCN/SSC CBSG 1992.

Supriatna, J.; Tilson, R. Javan gibbon and Javan langur action plan. Pp. 99-102 in JAVAN GIBBON AND JAVAN LANGUR POPULATION AND HABITAT VIABILITY ANALYSIS REPORT. J. Supriatna; R. Tilson; K.J. Gurmaya; J. Manansang; W. Wardojo; A. Sriyanto; A. Teare; K. Castle; U.S. Seal, eds. Apple Valley, Minnesota, IUCN/SSC CBSG, 1994.

Supriatna, J.; Tilson, R.; Gurmaya, K.J.; Manansang, J.; Wardojo, W.; Sriyanto, A.; Teare, A.; Castle, K.; Seal, U.S., eds. JAVAN GIBBON AND JAVAN LANGUR POPULATION AND HABITAT VIABILITY ANALYSIS REPORT. Apple Valley, Minnesota, IUCN/SSC CBSG, 1994.

Tilson, R.; Seal, U.S. Problem statement. Pp.5-7 in THAI GIBBON POPULATION AND HABITAT VIABILITY ANALYSIS. S. Tunhikorn; W. Brockelman; R. Tilson; U. Nimmanheminda; P. Ratanakorn; R. Cook; A. Teare; K. Castle; U.S. Seal, eds. Apple Valley, Minnesota, IUCN/SSC CBSG, 1994.

Woodruff, D.; Tilson, R. Genetic aspects of gibbon management in Thailand. Pp. 43-45 in THAI GIBBON POPULATION AND HABITAT VIABILITY ANALYSIS. S. Tunhikorn; W. Brockelman; R. Tilson; U. Nimmanheminda; P. Ratanakorn; R. Cook; A. Teare; K. Castle; U.S. Seal, eds. Apple Valley, Minnesota, IUCN/SSC CBSG, 1994.

Dr. Ronald Tilson is the Director of Conservation at the Minnesota Zoo and Coordinator of the Gibbon Species Survival Plan© for the AZA. He is also the Southeast Asian Programs Coordinator for the IUCN Conservation Breeding Specialist Group (CBSG) and has organized Population and Habitat Viability Assessments for Thai gibbons, Javan gibbons and langurs, and orangutans - among other mammal and bird species. **Katherine Castle** serves as the Co-Coordinator of the AZA Gibbon Species Survival Plan and was formerly the Curator/Tropics Mammals at the Minnesota Zoo. She is currently with the Minnesota Department of Natural Resources. **Dr. Jatna Supriatna** is with the University of Indonesia and **Kunkun Jaka Furmaya** is with the University of Padjaran. **Dr. Warren Brockelman** has done extensive field work with *Hylobates lar* and works with the Center of Conservation Biology at Mahidol University. **Schwann Tunhikorn** is with the Wildlife Research Division of the Thai Royal Forest Department.

An 18 year old silverback male mountain gorilla (*Gorilla gorilla beringei*), one of the study subjects at Karisoke Research Center. There are only 600 members of this subspecies remaining on Earth. (Photo by Dieter Steklis.)

PARTNERS IN CONSERVATION: ESTABLISHING *IN SITU* PARTNERSHIPS TO AID MOUNTAIN GORILLAS AND PEOPLE IN RANGE COUNTRIES

Charlene Jendry
Columbus Zoological Gardens, Powell, Ohio

INTRODUCTION

In working to provide state-of-the-art care for captive animal populations, many zoos have benefited from information provided by field researchers. For example, the pioneering work of Schaller [1963], Fossey [1974, 1984], and their colleagues produced a wealth of knowledge on the behavior and ecology of mountain gorillas (*Gorilla gorilla beringei*). Many zoos have adapted this information to supplement husbandry techniques for captive lowland gorillas *(Gorilla gorilla gorilla)*. For example, during a visit to the Columbus Zoo, Dian Fossey spent several days talking with the staff about behavior of free-living gorillas. Among other things, she discussed the apes' practice of building sleeping nests. Upon Fossey's advice, the gorillas were provided bales of hay, which they immediately used to construct nests similar to those found in African forests.

Zoos are now exploring ways to assist *in situ* conservation and research. The Wildlife Conservation Society and the Roger Williams Park Zoo [Savage et al., this volume] are examples of zoological organizations – large and small – that have effectively supported and conducted *in situ* projects. As *in situ* involvement has increased, so has its focus. Rather than limiting support to research and conservation of animals, such efforts are now involving projects that also benefit the lives of human populations. The conservation movement has evolved and become multi-faceted; primary components now include research, education, communication, fund-raising and the aid of indigenous people. Illustrating this approach, the present chapter provides a review of Partners In Conservation (PIC), a program established at the Columbus Zoo, in Ohio.

Following newspaper and television accounts of the outbreak of civil war in Rwanda in 1991, Partners In Conservation was founded by staff and docents at the Columbus Zoo. Because the Zoo's gorilla socialization and propagation program benefited by adapting information reported by field researchers in Rwanda, a program designed to assist free-living mountain gorillas and the indigenous people in

Figure 1. Roz Carr poses with children at the Imbabazi Orphanage in Rwanda. In 1994, Ms. Carr founded the orphanage to assist children whose parents were killed during the civil war. Columbus Zoo's Partners in Conservation program makes annual donations to the orphanage. (Photo courtesy Roz Carr).

range countries would integrate the *ex situ* and *in situ* conservation goals of the Columbus Zoo.

PIC's goal is to help preserve the future of the mountain gorilla and to improve the lives of indigenous people in the three range countries of Rwanda, Uganda, and Zaire. PIC objectives to achieve this goal are: (1) to develop and implement diverse educational programs that increase public awareness in the U.S. about the people and the endangered mountain gorillas of Rwanda, Uganda, and Zaire; (2) to raise money to fund anti-poaching patrols, gorilla behavior research, and veterinary care for the mountain gorillas; and (3) to identify and fund programs that benefit the people of Rwanda, Uganda, and Zaire.

With financial support from the Columbus Zoo, this grass-roots program works within a collaborative environment. PIC blends the knowledge of zoo staff, docents, educators, researchers, and the business community – primarily on a volunteer basis – to achieve its goal. By collaborating with a variety of groups, PIC is establishing partnerships among the public, private and nonprofit sectors.

PIC was organized using a model that the business community found to be successful. As Watson [1994] stated, "Since the private sector has developed into a knowledge of specialists, it has to be an organization of equals. No knowledge ranks higher than another; each is judged by its contribution to the common task rather than by any inherent superiority or inferiority. Therefore, the modern organization cannot be an organization of boss and subordinate. It must be organized as a team" [p. 11]. From its inception, PIC professional volunteers have been part of a diverse team and have played an active role in program development and implementation. We believe that if individuals are encouraged and given the authority to make suggestions or changes, the program will grow exponentially. Using this approach, PIC benefits from the talents of people from a variety of professions and, in turn, the volunteers benefit from their ownership in the program.

IN-SITU PARTNERSHIPS

The Fossey Fund. To aid the conservation of mountain gorillas, PIC initiated a partnership with the Dian Fossey Gorilla Fund (Fossey Fund) in 1991. PIC partially funded the repair of buildings at the Karisoke Research Center, damaged during the Rwandan civil war, and helped pay for anti-poaching patrols. The Columbus Zoo and the Fossey Fund agreed that, by collaborating through Partners In Conservation, they would expedite the conservation goals of both organizations. When zoos collaborate with *in situ* projects, it enhances the utility of efforts and demonstrates the important role zoos must play in global conservation [Amato, 1995].

Figure 2. Veterinarian Suzanne Anderson provides an antibiotic injection for a wild mountain gorilla in Rwanda. (Photo by Ruth Keesling.)

The Imbabazi Orphanage. In 1994, the Imbabazi Orphanage was founded in Rwanda to help children whose parents were killed during the civil war (Figure 1). Communication between PIC and the Director and founder of Imbabazi, Ms. Rosamond Carr, identified a need for financial assistance to operate the orphanage. In particular, funds were needed for the salaries of 24 Rwandans hired to care for the 70 orphaned children. Because this project fit one of PIC's objectives of assisting people living in mountain gorilla habitat countries, PIC pledged to fund $10,000 annually for the operating expenses of the Imbabazi Orphanage. Funding began in 1995; some of the fund-raising projects are described below.

The Mountain Gorilla Veterinary Project. PIC began another *in situ* partnership in 1995 with the Morris Animal Foundation's Mountain Gorilla Veterinary Project (MGVP). The MGVP, established in 1986, provides field veterinary care for mountain gorillas (Figure 2). MGVP veterinarians have successfully treated snare injuries and have managed several severe outbreaks of diseases among the gorillas. PIC initiated a partnership with the MGVP by contacting their Project Sponsor and Director. Discussions identified a need for improved, locally-available medical care for the gorillas. PIC awarded the MGVP a $5000 grant that helped to provide on-site advanced training of an African veterinarian, who received his veterinary degree in Zaire. His enhanced field veterinary skills will supplement the available expert medical care, perhaps aiding the long-term survival of the mountain gorillas.

Rwandan Artisan Project. In 1996, using the in-country contacts of PIC, the Fossey Fund and MGVP, a "People to People" partnership was initiated with indigenous artisans living adjacent to the gorillas' habitat in the Virunga Mountains of Rwanda. In cooperation with PIC, the MGVP Director, the late Dr. Jim Foster, presented the following plan to local artisans:

- PIC proposed the formation of a partnership with local (Rwandan) artisans, whereby PIC would purchase art made in Rwanda and sell it in the U.S. to raise funds for conservation;
- The program will be administered in Rwanda, by Rwandans;
- PIC agrees to pay a fair price for products and will not ask artisans to make items that are not created locally; and
- The products will be sold by PIC and all proceeds will benefit gorillas and people in PIC partnership projects.

The group received the plan as presented and voted to name the project *Terimbere Uzigamiru Kinigi*, or "The Conservation and Development of Kinigi." (Kinigi is the village near park headquarters for the *Parc National des Volcans*.) A Rwandan was selected to serve as the coordinator for the program and, in the spring of 1996, sample products from 15 artisans were purchased for test marketing in the United States.

In October, 1996, the author met with the artisan program's coordinator and artists in Kinigi. Two hundred hand-crafted items were purchased from 18 individuals and from two women's cooperatives. The merchandise, including masks, nativity sets, and dolls, were transported to the United States by the author. Ninety-

Figure 3. Elementary students in Columbus, Ohio, carefully examine snares collected by anti-poaching patrols at the Karisoke Research Center. (Photo by Charlene Jendry.)

eight percent of the merchandise was purchased during the first week of sales. Another order was placed with the artisan coordinator in January of 1997. The three orders provided $3,000 income to the artisans and $3,900 to benefit PIC partnership projects in Rwanda and Uganda.

Because the infrastructure for shipping to and from Rwanda has not been consistent following the civil war, merchandise are hand-carried out of Rwanda by researchers and missionaries. The artisan products are sold at zoos, at PIC adult programs (described below), and through mail orders. Future plans include providing tools, supplies and a small building in which the artisans may make and store their merchandise.

EDUCATIONAL PROGRAMS FOR STUDENTS IN THE U. S.

Many of today's conservation education initiatives target children as their audience. While it is true that students have a limited ability to make major efforts toward conservation, there are many small projects (described below) that can raise awareness and influence change. Moreover, today's children will one day be voters, consumers, international tourists, active citizens of the world. If they learn today, they may act tomorrow. The PIC philosophy is that early education can gently shape life-long attitudes about conservation, ecology, and benevolent care toward fellow humans.

To address this issue, PIC initiated the IN-SCHOOL Program in 1993 for students, aged Kindergarten to 12th grade (K-12). PIC piloted the IN-SCHOOL Program in seven schools in Columbus, Ohio. Teachers who participated in the pilot program became the nucleus for a PIC Teacher Advisory Board. The IN-SCHOOL program has expanded and, by 1996, included students from 36 schools in Ohio, New York, and Illinois. The program is presented free-of-charge at inner-city and suburban schools - both public and private.

To understand better any conservation issue, it is important for students to be exposed to information about the people and the history of range countries. It is difficult for students to be interested in other countries unless they feel that their own futures are interconnected with the future of others [Martin, 1994]. Therefore, the objective of the IN-SCHOOL Program is to present conservation programs that focus on the people, cultures, and mountain gorillas of Rwanda. In the future, PIC will also incorporate cultural information relating to the people of Uganda and Zaire, to provide a broader multi-cultural curriculum.

Creative education initiatives are critical components in all areas of PIC programming, because they are the foundation for teaching, processing and internalizing the conservation message. To assure the success of the IN-SCHOOL Program, PIC organizers and educators designed a four-step plan, outlined below.

Figure 4. Trackers and staff at the Karisoke Research Center are wearing "friendship bracelets" made by children in PIC's IN-SCHOOL Program. (Photo by Peter Clay.)

1.) **Teacher Orientation Workshops.** PIC co-founders and the Teacher Advisory Board conduct an orientation workshop for teachers. Information about the people of Rwanda, rainforest preservation, mountain gorilla behavior and conservation is intertwined into a multi-disciplinary curriculum for the teachers who, in turn, volunteer their time and lend their expertise to the program by adding further to the curriculum. The teachers have integrated information provided by PIC into existing lesson plans or designed new ones in 26 subjects including math, art, ecology, biology, geography, economics, public speaking, music, creative writing, reading and foreign languages. After the workshop, teachers complete a program evaluation survey, thereby providing helpful feedback to the Teacher Advisory Board in modifying and improving future workshop presentations.

2.) **PIC Educational Lectures for Students.** Following classroom instruction, PIC representatives present an educational lecture that includes slides and a narrated video featuring the people and mountain gorillas of Rwanda. After the lecture, students are invited to examine Rwandan artifacts. We demonstrate how snares work (Figure 3) and also give students an opportunity to use an *ingata*, play a flute made by a tracker at Karisoke, and examine Rwandan toys. These hands-on activities enable students, ages five through 18, to learn first hand about Rwandan cultures.

3.) **Action-oriented Conservation Projects.** PIC programs focus on experiential learning, rather than using a passive/receptive process. Therefore, the IN-SCHOOL Program requires that students engage in action-oriented projects designed to help them learn new conservation concepts. More than 20,000 students have completed projects during the IN-SCHOOL Program. Examples of student projects are described below.

- To integrate cultural information into the curriculum, teachers taught students how to make and use fabric *ingatas*, circular objects traditionally made by braiding fabric or grass. *Ingatas* are worn on the heads of Rwandans to help balance food, water, or firewood. Since most Rwandans do not travel by automobile, this activity enabled students to experience how Rwandans (as well as people in other parts of the world) transport items from place to place.
- Children with learning deficits worked with their teacher to construct three-quarter to scale *papier mache* models of an adult female, infant, and silverback male gorilla. The exercise helped children learn about size differences and development. The lesson also included information about behavior and social structure of gorillas.
- Teachers accompanied elementary students to a field near their school and taught them how to construct nests similar to those used by gorillas. To illustrate gorilla foraging behavior, teachers helped the students look for popcorn hidden in the field.

- A third grade teacher asked students to research rainforests and their inhabitants. The class then used computer graphics to illustrate a presentation discussing rainforests of the world.
- A home economics class designed and made a quilt with a conservation theme.
- Classes participated in recycling projects associated with classroom studies; students elected to donate the proceeds to the Imbabazi Orphanage.
- Students presented educational puppet shows, wrote letters to children in Rwanda and displayed posters in their communities depicting both the people and gorillas of Rwanda.
- Elementary classes conducted "Gorilla School" for their parents, Columbus School Board Members, Ohio State senators and representatives of the Columbus Zoo's Board of Trustees. The children serve as "teachers", demonstrating what they had learned in the IN-SCHOOL Program.
- Middle school students wrote and performed a "Gorilla Rap."
- In April of 1995, students in the program created over 5,000 "friendship" bracelets made by braiding yarn. The bracelets were given to children at the Imbabazi Orphanage and to Rwandans of all ages who live near the Parc National des Volcans (Figure 4). The bracelets were presented with the message that they were made by children in PIC's IN-SCHOOL Programs in the United States. The Rwandans responded with appreciation for the small gifts, saying they were unaware that people outside Rwanda knew about the war or cared about Rwandan citizens. The message of mutual appreciation came full circle when PIC representatives showed photographs of Rwandans wearing the friendship bracelets to the U.S. students who made them.
- In 1995, PIC invited IN-SCHOOL participants to design posters on various aspects of conservation. Teachers selected three posters from each school and these were matted by PIC volunteers. The students' work was the centerpiece of an art exhibit entitled, "Conservation: A Student Perspective." The exhibit was displayed at the Columbus Zoo during the summer of 1995 and, in September of that year, was featured at an art gallery in Columbus, Ohio.
- In 1996, students made over 2,000 fabric teddy bears created from recycled materials. Some of the bears were distributed to Rwandan children who live near the Parc National des Volcans. Other bears were distributed closer to home, at a Columbus, Ohio, battered women's shelter and to children admitted to local hospital emergency rooms. The focus of this project was to give children in the IN-SCHOOL Program an opportunity to do something to help others - both in another country and in their own community.

4.) **Followup.** After students complete their projects, a PIC representative makes a second visit to each school. A certificate of merit is presented to the school, acknowledging the collective efforts of students in aiding the mountain gorillas and the people of Rwanda. This second visit to the schools also allows students to demonstrate their projects and ask questions generated since the first visit of the PIC representative.

Prior to the IN-SCHOOL Program, most students could not locate Rwanda on a world map, did not know which languages were spoken in Rwanda, nor did they know basic information about mountain gorillas such as their social structure, vocalizations, or conservation status. When tested after the program, however, most students could correctly answer questions relating to these issues.

Teacher evaluations of the IN-SCHOOL Program determined that students K-12 successfully assimilated information about mountain gorillas and developed an understanding and appreciation for another culture. Included in the evaluation forms were additional comments regarding: enhanced student motivation, increased vocabularies, and greater student cooperation while working on the IN-SCHOOL Program.

Partners In Conservation's IN-SCHOOL Program was designed to be used in part or in its entirety by any zoo or other organization. In 1995, an Advisory Team was formed to help other groups to implement the program. Thus far, PIC has assisted the following:

- At the invitation of Girl Scout leaders in Columbus, the IN-SCHOOL Program was incorporated into the scouting schedule in 1995. The Advisory Team invited cadet scouts to help design interactive conservation projects that met scouting requirements.
- In 1996, the Pittsburgh Zoo received a grant from Blue Cross of Western Pennsylvania to implement PIC's IN-SCHOOL Program. The Advisory Team conducted two workshops for members of the Education Staff at the Pittsburgh Zoo. Teaching materials included snares, *ingatas*, slides, videos and a complete curriculum. The Pittsburgh Zoo will begin the IN-SCHOOL Program in 1997.
- In their "Five-Year Strategic Plan - 1995", the Bonobo Species Survival Plan© (SSP) recommended using the model initiated by Partners In Conservation to develop school programs for bonobo *(Pan paniscus*; pygmy chimpanzee) conservation education [Reinartz, 1995; and see Reinartz and Boese, this volume]. The PIC Advisory Team will work with the Bonobo SSP members to adapt the IN-SCHOOL Program.

EDUCATIONAL PROGRAMS FOR ADULTS IN THE U.S.

Knowles [1984] writes: "understanding about adult learning is an essential factor in effecting a change in adult thinking. Learning involves change, because it enables the individual to make both personal and social adjustments" [p. 6]. PIC incorporated Knowles's ideas when designing adult programs that feature a holistic approach to conservation. This message is conveyed through the personal experiences of the author and is illustrated through video and slides depicting mountain gorillas and images of various PIC projects. In taking this approach to an adult audience, PIC strives to give individuals a "visceral" experience. The program offers adults an opportunity to discuss their pre-conceptions about conservation, e.g., that it only relates to the protection of animals and their natural habitats.

PIC's adult programs endeavor to educate and to engage the audience in constructive dialogue. The program informs people that they can participate in a variety of conservation projects both in range countries and in their local communities. More than 30,000 people have attended PIC adult programs presented at professional meetings, service organizations, universities, churches, other zoos, and at international sites such as the Museum of Natural History in London, England. Many people who attend PIC adult programs become PIC volunteers and/or make donations to the program.

PUBLIC AWARENESS EFFORTS

Partners In Conservation conducted two non-political petition campaigns in 1991 and 1993, supporting a cease-fire in Rwanda. In 1994, a larger "Petition for Peace in Rwanda" was co-sponsored by PIC and the Fossey Fund, using the network system established in the first two campaigns. In only eight weeks, citizens from 50 countries and all 50 states in the U.S., collected signatures from people of diverse backgrounds. Hundreds of volunteers worked on behalf of zoos, museums, and universities to circulate the petitions in communities worldwide. To maximize the effectiveness of the petitions, the author talked with the Public Relations Officers for both the Secretary General of the United Nations, Boutros-Boutros Gahli, and Madeline Albright, United States Ambassador to the United Nations. Following those conversations, the petitions were requested by Madeline Albright, in July 1994. Ambassador Albright stated in a personal letter to the author, "I wish to thank your colleagues for their strong commitment to the people of Rwanda and for helping to focus on peace for this region." PIC also received letters from the Rwandan Ambassador to the United States and from the Director for Diplomatic Affairs for the Rwandan Patriotic Front. These two letters offered differing points of view concerning the war, but each expressed appreciation for the petition's effort to increase public awareness about Rwanda's plight.

Clearly, a petition cannot bring about the end of a war – regardless of how many signatures are obtained. However, by participating in petition campaigns, a quarter of a million people from the world community made a strong statement that they care about the future of Rwanda's people and wildlife. As a public awareness vehicle, the Petition for Peace Campaigns were a great success.

Nearly 40 reports highlighting PIC activities – including the IN-SCHOOL Program, international Petition for Peace Campaigns, and partnerships between the Columbus Zoo and *in situ* projects – have appeared in newspapers and on television in local, national, and international markets. This publicity has allowed PIC to reach an extended audience and the positive coverage has increased public awareness about the conservation initiatives of the Columbus Zoo, the Fossey Fund, the Mountain Gorilla Veterinary Project and efforts to help children at the Imbabazi Orphanage.

In 1994, the Columbus Zoo opened a multi-media exhibit to offer visitors an opportunity to learn about the zoo's collaboration with *in situ* projects around the world. Graphics and a narrated video describing PIC projects were highlighted. After seeing the conservation exhibit, many visitors requested further information on how they could become involved in the PIC program.

PIC FUND-RAISING

PIC's fund-raising objectives are being achieved through donations, sale of merchandise, special fund-raising events, and partnership projects. Because administrative costs are covered by the Columbus Zoo, all proceeds directly benefit PIC projects. To date, PIC has generated $115,000 by providing educational programs and offering people choices for financial involvement.

Private Donations through Lecture Programs. PIC co-founders believe that, if we effectively communicate a conservation message, people will want to join the effort to help. During the adult lecture series, we offer a menu of opportunities allowing people to volunteer in a project or contribute monetarily. Fund-raising has been a natural outgrowth of effective educational programming.

Merchandise. Partners In Conservation develops and markets its own merchandise. Zoo staff and docent artists provide original artwork and many items are designed and produced by creative volunteers. The merchandise is sold at PIC-sponsored adult lectures, at the Columbus Zoo gift shops, through mail orders and by staff and volunteers at other zoos. PIC began marketing one T-shirt design in 1991 and, in 1993, expanded to include a broader range of products, including originally designed jewelry and stationery.

Special Events. In addition to selling merchandise and obtaining donations, PIC has participated in several special events to generate funds for conservation. Two of the most successful events were:

- The Columbus Zoo sponsored a "Walk For Wildlife" on Earth Day, 1995, which netted $4,500. PIC provided and staffed an educational booth at the event. The funds were equally distributed by PIC to the Fossey Fund and the Imbabazi Orphanage in Rwanda.
- PIC and Ross Products Division of Abbott Laboratories hosted the first annual Rwandan Fête (or celebration) in 1995, honoring the life and work of the late primatologist, Dr. Dian Fossey. The Fête program was designed to increase awareness about the people and mountain gorillas of Rwanda

by immersing guests in Rwandan culture, including a sampling of traditional food and music. The 250 guests contributed over $20,000 that equally benefited the children at the Imbabazi Orphanage and the mountain gorillas through the Fossey Fund. The Second Annual Rwandan Fête, held in 1996, honored the humanitarian efforts of Ms. Rosamond Carr. The event generated $21,000, that was equally distributed to the Imbabazi Orphanage and the Mountain Gorilla Veterinary Project.

Partnership Projects. PIC has experienced fund-raising success through establishing partnerships with other organizations. These partnerships access the expertise of additional volunteers and new audiences to attend PIC's conservation lectures. Some of the partnership projects include:

- Merlin Marketing, Inc., of Atlanta, Georgia, became a business partner with PIC in 1993. The company marketed gift boxes depicting illustrations of endangered species. The PIC logo appeared on each box and Merlin Marketing donated 1% of the profit to PIC. A total of $835 was received from this campaign. (Note: $800 covers the operating expenses of the Imbabazi Orphanage for one month.)

- PIC approved a grant request from a member of Chi Omega Sorority at The Ohio State University in 1994. The student used a grant of $300 to purchase and refurbish five gumball machines. Graphics explaining PIC objectives were added to the dispensers. The gumball machines were then placed in five sorority houses. The project is supervised by the grant recipient and provides PIC with educational and fund-raising opportunities. Several hundred dollars is generated each year from this project. The gumball project led the grant recipient to establish UNITED, a chartered organization at The Ohio State University. The goal of UNITED is to encourage students to be involved in conservation projects through partnership with PIC. UNITED objectives are to present a PIC adult lecture program for Ohio State students each year and to initiate one fund-raising event per year. All proceeds will be donated to PIC. In 1996, UNITED presented a PIC lecture on campus and organized "BANANA-NANA", a benefit rock show that feature five local bands. The event generated over $900.

- PIC invited 16 zoos in the United States to join the Columbus Zoo in raising funds for the Imbabazi Orphanage. Eight zoos responded to the challenge and developed their own local fund-raising projects. From this collective effort, a check for $5,000 was sent to Ms. Carr at Imbabazi, with letters from each zoo describing how the money was raised. (This $5,000 was supplementary to PIC's $10,000 annual pledge to Imbabazi.)

- In 1995, High Chief Productions, headquartered in Columbus, Ohio, established a partnership with PIC in conjunction with the release of a compact disc, entitled *Jack Hanna's World*. The CD features original music by Emmy Award winning composer, Mark Frye. The PIC logo appears on the cover and a page in the insert booklet explains PIC's mission. PIC receives $.25 from the sale of each CD, distributed nationally and internationally by Virgin Records. This partnership provides PIC with an excellent opportunity for educational and financial rewards.

WHY DOES PIC WORK?

In a very short time, Partners In Conservation has become an established program, known for its innovative and energetic efforts to accomplish its conservation education goals. Key factors playing a role in PIC's success are:

Cooperation. PIC is cooperatively administered by the four founding members - Barb DeLorme, Judy Hoffman, Jeff Ramsey, and the author. All decisions are made jointly and each co-founder brings to the program specific areas of expertise, including marketing, accounting, graphics design, and public relations. PIC blends the professional backgrounds of its co-founders with the specialized talents of volunteers to carry out its objectives. Cooperative networking between individuals, zoos, businesses and *in situ* projects has been an important factor in PIC's success.

Communication. Initially, PIC co-founders used established professional and business contacts to enlist volunteers. After the first year, however, individuals and businesses began to approach PIC to explore how they could become involved. Word of mouth, as well as strong and positive media coverage, has helped to generate a wide and active volunteer base.

Flexibility. PIC does not have membership dues nor regularly scheduled volunteer meetings; volunteers are contacted when their talents are needed. Not having required meetings provides a flexible system which accommodates the volunteers' schedules. This approach requires careful organization on the part of the co-founders, but benefits PIC by increasing the pool of professionals willing to offer their time and energy to the program.

Organization. The sale of PIC merchandise is operated like a small business. Prior to 1994, the Columbus Zoo paid for the merchandise and, following the sale of merchandise, PIC reimbursed the zoo before earning a profit. Since 1994, PIC has paid for the merchandise and now owns its entire inventory. Because the merchandise has a rapid turnover, investment cost is kept low. The Columbus Zoo provides both the author's salary and office expenses for PIC, allowing all proceeds from fund-raising activities to go directly to PIC projects.

CONCLUSIONS

Partners In Conservation, based at the Columbus Zoo, is meeting its goal to support *in situ* conservation of mountain gorillas and to assist the people of the Virunga Mountains in Rwanda. PIC blends the talents of zoo staff and volunteers and fosters alliances between zoos, field researchers, conservation organizations, educators and the business community. Because adults and students actively participate in Partners In Conservation projects, they become an integral part of the effort. This "people connection" is the component that provides PIC with a diversity of ideas and the people to put these ideas in motion. Among PIC's accomplishments:

- Petition campaigns enabled over one-quarter of a million people from around the world to voice their concern for the people and wildlife in Rwanda.
- The IN-SCHOOL Program blends the knowledge of educators and the zoo community to create programs encouraging involvement of students in conservation. More than 20,000 students have completed the program.
- The adult program gives individuals information about the people and mountain gorillas of Rwanda and encourages them to adapt a holistic approach to conservation.
- Fund-raising activities have enabled PIC to donate money to support the work of the Dian Fossey Gorilla Fund at the Karisoke Research Center.
- PIC donated to Doctors Without Borders - Rwandan Relief.
- The alliance between PIC and the Imbabazi Orphanage allows for concerned U.S. citizens to help children whose parents were killed during the war in Rwanda.
- The partnership between PIC and the Mountain Gorilla Veterinary Project enabled PIC to fund the advanced training of an African veterinarian. By doing so, PIC assured improved veterinary support for mountain gorillas.
- PIC donated funds to the Mountain Gorilla Veterinary Project in Uganda for the purchase of medical books for veterinary students at the University of Makerere, Kampala, Uganda.
- PIC's participation in the Columbus Zoo's conservation center informed the community of the Zoo's commitment to *in situ* conservation and the people in range countries.

Partners In Conservation is one example of how zoos and aquariums of any size can advance their conservation goals. The program demonstrates that zoos can effectively collaborate with *in situ* projects to assist both endangered wildlife and indigenous people in range countries. In addition, by focusing on the educational factor in the conservation message, Partners In Conservation shows that people of all ages, all over the world, can be empowered to work together, form partnerships, raise money and contribute to primate conservation.

ACKNOWLEDGMENTS

The success of Partners In Conservation is due in large part to the creativity, commitment, and hard work of Barbara deLorme, Judy Hoffman, and Jeff Ramsey. The four of us wish to thank the Columbus Zoo for their encouragement and continued support - both philosophically and financially. Partners In Conservation's growth and vitality can be directly traced to the tireless efforts of the Zoo's staff and docents and to the educators who collaborate with us on the IN-SCHOOL Program.

In addition, we want to express our gratitude to Dr. Jim Foster, Project Director for the Mountain Gorilla Veterinary Project, who was instrumental in the establishment of the Rwandan Artisan Program. Dr. Foster died May 10, 1997. His gentle nature, good advice and friendship will be missed by all who knew him.

REFERENCES

Amato, G. Zoos as centers for conservation biology. Pp. 205-208 in PROCEEDINGS OF THE AMERICAN ZOOLOGICAL ASSOCIATION REGIONAL CONFERENCE. Wheeling, West Virginia, American Zoological Association, 1995.

Fossey, D. Observations on the home range of one group of mountain gorillas (*Gorilla gorilla beringei*). ANIMAL BEHAVIOUR 22:568-581, 1974.

Fossey, D. Infanticide in mountain gorillas (*Gorilla gorilla beringei*) with comparative notes on chimpanzees. Pp. 217-235 in INFANTICIDE: COMPARATIVE AND EVOLUTIONARY PERSPECTIVES. G. Hausfater; S.B. Hrdy, eds. New York, Aldine, 1984.

Knowles, M. THE ADULT LEARNER: A NEGLECTED SPECIES. 4th Ed., Houston, Texas, Gulf Publishing Company, 1993.

Martin, J. Who cares about Africa? AMERICA 172(17):16-20, 1995.

Reinartz, G. SSP: BONOBO FIVE-YEAR STRATEGIC PLAN - 1995. Milwaukee, Society of Milwaukee County Zoo, 1995.

Schaller, G.B. THE MOUNTAIN GORILLA: ECOLOGY AND BEHAVIOR. Chicago, The University of Chicago Press, 1963.

Watson, G. BUSINESS SYSTEMS ENGINEERING: MANAGING BREAKTHROUGH CHANGES FOR PRODUCTIVITY AND PROFIT. New York, John Wiley and Sons, Inc., 1994.

Charlene Jendry serves as Outreach Coordinator for the Columbus Zoo, where she initiated Partners In Conservation in 1991. She is also a member of the Gorilla SSP Behavioral Advisory Group and the AZA's new East Africa Faunal Interest Group. Charlene is the editor of the *Gorilla Gazette*, an international newsletter dedicated to the exchange of information between people working with captive gorillas and those working with wild populations. She has worked with captive gorillas for 14 years and made a brief study visit to the Karisoke Research Center in Rwanda.

This bonobo, or pygmy chimpanzee (*Pan paniscus*), belongs to the rarest species of great ape. Their native habitat is confined to Zaire. At present, there are just over 100 bonobos in captivity worldwide. (Photo by Richard Bartz.)

BONOBO CONSERVATION:
THE EVOLUTION OF A ZOOLOGICAL SOCIETY PROGRAM

Gay E. Reinartz and Gilbert K. Boese
Zoological Society of Milwaukee County, Wisconsin

INTRODUCTION

Zoos today are expected to engage in wildlife conservation efforts. The concern over vanishing wildlife and degradation of the environment has led to a dramatic shift in the way society views wildlife. We are urging public institutions to take an active role in protecting nature. In response, zoos are developing conservation activities that go far beyond their traditional roles, beyond their locale, and beyond captive breeding [Koontz, this volume].

In many cases, the shift toward a conservation mission has required zoos to build a new infrastructure and funding base from which to include conservation initiatives. Effective conservation programs must have stable funding sources and a long-term commitment. However, municipally-owned zoos are often subject to budgetary constraints due to competition with other government programs and perhaps government indifference. When government officials are unable to commit financially to wildlife conservation, zoos can develop partnerships with private groups to accomplish their goals. As an example, the Zoological Society of Milwaukee County (ZSMC) is a non-profit membership organization which partially supports the Milwaukee County Zoo. In addition to contributing to the Zoo's operating funds, the ZSMC supports conservation programs, funded by the private sector and administered by the ZSMC. Since 1990, the ZSMC has expanded its program orientation from local zoo support to include international conservation initiatives. The ZSMC and Milwaukee County Zoo's public-private partnership serves as a model of how zoo-based organizations can contribute to primate conservation. While ZSMC's conservation programs encompass a variety of endangered species and habitats, for the purpose of this discussion we focus on the captive breeding and conservation of bonobos (*Pan paniscus*). We describe the conservation status of bonobos and the evolution of ZSMC's involvement with this species, starting as managers of the bonobo captive breeding program and developing as supporters of field conservation projects.

THE BONOBO: ENDANGERED GREAT APE

The bonobo, or pygmy chimpanzee, has been described as the rarest of all great ape species [Oates, 1985]. The species occurs in a relatively small forest region of Zaire*, south of the Zaire River in the central river basin. As far as we know, it is allopatric to chimpanzees (*Pan troglodytes*) which occur north of the river. Bonobos are found predominantly in rain forest habitat. However, recent reports of a resident population living in a savanna-mosaic forest expands their ecological boundaries and confirms what was once considered the southern extent of their range [J. Thompson, personal communication; Thompson-Handler et al., 1995].

As seen in all great ape species, habitat destruction and increasing human population pressure are among the greatest threats to bonobo survival [Malenky et al., 1989; Susman et al., 1981]. The rain forest of Zaire is one of the largest unspoiled forests remaining in Africa, but it also represents potential economic wealth to a country whose human needs are great. Because of the forest, the country harbors a vast array of endemic species; Zaire is ranked fourth in the world for its level of biodiversity [McNeely et al., 1990]. The remoteness of this forest region has delayed – but not stopped – industrial deforestation, conversion of forests for agricultural use, and human encroachment. These factors are primary agents of fragmentation and decline of wild bonobo populations [Thompson-Handler et al., 1995]. Bonobos are also directly exploited by hunters and poachers and once fetched high prices through international trade. As recently as 1990, an adult pair (held in captivity) was sold for over $250,000 to a non-U.S., non-AZA zoo. In parts of their range, bonobos are hunted for meat. Even in areas where there is a strong cultural taboo against eating apes, these taboos break down as human populations migrate from urban areas to seek subsistence living or flee ethnic violence [Kano et al., 1994; Tashiro, 1995]. Furthermore, commercial logging is gradually making remote areas more accessible. As a result, over the past few years, field biologists have reported an increase in bonobo poaching concomitant with an alarming increase in the bush meat trade [see "Site Reports" in Thompson-Handler et al., 1995].

There are no recent census data available for the bonobo in its present range. The species no longer exists in much of its historical range, and existing populations are highly fragmented [Susman et al., 1981; Susman & Kabongo, 1984]. Several research sites have experienced a dramatic drop in population size over the past decade [Kano et al., 1994]. Researchers estimate that fewer than 25,000 bonobos remain in the wild, and more likely the number is between 10,000 and 20,000 [Thompson-Handler et al., 1995]. Although this may appear to be a large number, factors such as long interbirth interval and long adolescent infertility result in slow population growth, making this species vulnerable to environmen-

---

* Although the country was renamed the "Democratic Republic of the Congo" in May, 1997, the name "Zaire" is used throughout this chapter.

tal pressures. In addition, bonobos are endemic to an area within one political boundary, placing them in a much more precarious position than if their range was spread across several countries.

THE SSP: A MANAGEMENT PLAN FOR SPECIES SURVIVAL

In 1987, the World Conservation Union (IUCN) issued the following policy statement:

> IUCN encourages those national and international organizations and those individual institutions concerned with maintaining wild animals in captivity to commit themselves to a general policy of developing demographically self-sustaining captive populations of endangered species wherever necessary.
>
> Captive populations need to be founded and managed according to sound scientific principles for the primary purpose of securing the survival of the species through stable, self-sustaining captive populations. Stable captive populations preserve the options of reintroduction and/or supplementation of wild populations. [World Conservation Union (IUCN), 1987, pp. 2-3]

Thus, management of an *ex situ* population may serve as a supportive intervention for a species at high risk. Captive breeding of endangered species may provide some insurance against total extinction [Foose et al., 1987], even though for many endangered species (including bonobos) reintroduction of captive-reared individuals into the wild is currently untenable. Despite the small probability for successful reintroduction, the process of captive breeding yields ancillary benefits to species preservation through: 1) conservation education to heighten public awareness; 2) scientific insight from *ex situ* research; 3) advances in animal husbandry and veterinary science; and 4) direct access to additional resources and sources of funding for the conservation of wild populations.

The deterioration of wild great ape populations prompted review of their species status in captivity. Prior to the organization of captive breeding programs, there existed a variety of individual zoo animal collections that had little or no inter-connection. In the past, animal management had an institutional focus rather than a population focus, i.e., there was little attempt to provide for the welfare of a species' population as a whole over that of maintaining good individual exhibits. For example, there was no mechanism for maintaining genetic variability or setting standards for husbandry practices. Zoo populations became highly inbred and experienced local extinctions [Ralls & Ballou, 1983]; many species of primates showed a low tolerance for inbreeding [Ralls & Ballou, 1982]. In 1981, the American Zoo and Aquarium Association introduced the endangered species breeding and management program, the Species Survival Plan© or SSP [see Wiese &

Hutchins, this volume]. With the advent of this program, zoos changed their perspective. The greatest single impact of SSPs has been the shaping of zoo managers' treatment of animals as part of a larger picture, thereby diminishing the "collector" mentality characteristic of earlier times. Individual animals now function as members of a system-wide population, serving the greater purpose as a conservation resource.

All great ape species in North American zoos are represented by an SSP. Analogous programs, such as the European Endangered Species Programme (EEP) exist in Europe, and in other regions of the world. The goals of SSPs are multifaceted, but ultimately they are a consortia that manages a given species to form a long-term, self-sustaining captive population that will complement the conservation needs of the wild population. How this translates into practice and specific management objectives varies with each taxon, depending on its captive and wild status, biological and behavioral systems, and conservation needs. The Bonobo SSP was officially recognized in 1988. From the beginning, the ZSMC has supported all aspects of the Bonobo SSP — through the acquisition of grants from private foundations and the sponsorship of cooperative conservation projects.

Bonobos are rare in nature (Kano, 1992) and also in captivity. As of June, 1995, there were 107 (4 unconfirmed) bonobos in zoological institutions worldwide (outside Africa); only 53 are in the SSP (as residents of the U.S. and Mexico). In comparison, there are approximately 350 chimpanzees, 300 western lowland gorillas, and 250 orangutans in U.S. zoos alone. Thus, in contrast to the management objectives for the other great ape species, it is clear that careful demographic and genetic management is necessary to create a self-sustaining captive bonobo population. This is possible only if all the animals are intensively managed for preservation of genetic diversity, which also requires attention to the social needs of the individual groups. In addition, we must have frequent exchanges between the SSP and EEP, which have to be managed as a single population, i.e., there must be gene flow between the two sub-populations to assure genetic diversity.

For great ape SSPs, reintroduction is not an immediate goal and may never be feasible given the complexities of reintroducing any species and the paucity of models for great ape rehabilitation and reintroduction. Even so, genetic management of bonobos serves to protect the captive population from potential loss of fitness that may be correlated with loss of genetic diversity from high levels of inbreeding. Although management objectives include selective breeding to minimize loss of genetic variability, the mission of the Bonobo SSP is not limited to population maintenance. The Bonobo SSP provides an example of how management programs direct the attention of the zoo community toward species preservation. The formation of the SSPs created the organizational structure for project collaboration among zoos, for both *ex situ* and *in situ* projects. The goals of the Bonobo SSP, as described by Reinartz [1994] and the Bonobo SSP Five Year Strategic Plan, are as follows:

Ex situ:

- Establish a self-sustaining captive population for *Pan paniscus* (per IUCN), coordinating breeding recommendations with the EEP in order to connect these regional gene pools.
- Utilize bonobos in zoos as an educational resource and as conservation ambassadors for their wild counterparts.
- Increase public awareness and appreciation for the species' uniqueness and conservation status.
- Sponsor educational activities and programs.
- Optimize the quality of life for captive bonobos - promote natural social groups and behavior.
- Support research applied to the conservation biology of the species.
- Provide training and technical assistance to facilities in less developed countries.

In situ:

- Using captive bonobos as ambassadors, use the collective ability of the SSP to raise funds for conservation work.
- Support conservation projects that benefit preservation of wild populations (as identified in an action plan).
- Support habitat protection and population monitoring.
- Support grassroots initiatives and local community involvement.
- Support conservation education and outreach projects.
- Promote the development of a holistic action plan for bonobos.
- Encourage and support research applied to the conservation biology of *Pan paniscus*.

HISTORY OF THE BONOBO SSP

In 1987, there were only 29 bonobos in the United States. Pedigree analyses indicated that the U.S. and the European sub-populations were essentially separate demes, each with a very limited gene pool [Reinartz, 1987]. Few institutions had a group size and/or composition resembling that of natural social formations and a significant number of offspring were human-reared without the early opportunity to learn normal, species-specific behaviors. These factors, combined with high infant mortality, impeded population growth. Furthermore, many facilities were of an older design [see Gold, this volume], inhibiting expansion of group size and limiting both the quality of life for the bonobos and the education value for the public. Thus, for the first five years, the Bonobo SSP focused on the imminent challenge of devising genetic and demographic management strategies for the precariously small captive population, understanding and addressing the social needs of the population, and improving the captive environment. These issues were seen as the first step toward addressing preservation of bonobos, however, the full conservation potential of the SSP did not develop until firm links to conservation in Zaire were established.

The extent of bonobo *in situ* conservation activities at that time revolved around initiatives connected with academic research and protection of study sites. Thus, *in situ* links for the Bonobo SSP were formed by establishing ties with field biologists. Beginning in 1988, Nancy Thompson-Handler, then Co-director of the Lomako Forest Pygmy Chimpanzee Project in Zaire, agreed to serve as the SSP's field liaison. This marked the beginning of an effort to bridge the gap between field and zoo conservation. In 1989, George Rabb [1990], Chairman of the Species Survival Commission, IUCN, challenged the SSP and the EEP to support a comprehensive action plan for preserving bonobos in Zaire, and to become involved in *in situ* conservation as one component of that holistic plan. From 1990 to 1992, the SSP participated in several international forums that gathered field researchers, Zairian wildlife officials, captive breeding specialists and related conservation organizations into working groups [Thompson-Handler et al., 1995; also see Konstant, 1992]. While the composition and nature of these groups were transitional, they marked the initiation of the Bonobo SSP and ZSMC into the network of bonobo conservationists. Discussions led to the identification of several conservation priorities:

- Survey specific areas in Zaire to more fully determine the distribution and abundance of bonobos;
- Continue to support research at established sites and expand research efforts to areas of critical ecological interest;
- Develop reserves and protected areas;
- Conduct education and conservation awareness programs that discourage "pity buying" in urban markets and hunting in rural districts;
- Train and promote professional development of Zairian conservation biologists.

Furthermore, the workshops reinforced the need for a bonobo conservation action plan. In 1991, Geza Teleki, of the (then) Committee for the Conservation and Care of Chimpanzees (CCCC), called for development of a bonobo action plan that would complement and follow the structure of the drafted chimpanzee action plan. The first publication was to consist of site reports from field investigators and a comprehensive review of all pertinent information regarding the decline and conservation of bonobos .

RECENT PROGRESS IN BONOBO CONSERVATION

Political and economic instability in Zaire has impeded large-scale work toward the listed goals. Therefore, most work sponsored by the ZSMC and the SSP has been conducted at the grassroots level. Funding for these projects come from private foundation grants obtained by the ZSMC and collaboration among SSP institutions.

Field Conservation Grants

Our conservation activities have depended almost entirely on relationships developed with field researchers. For the past three years, the ZSMC and the

Columbus Zoo have awarded conservation grants to students studying basic bonobo ecology. Not only does further research provide insight into the ecological parameters of this species, but the researchers have an investment in – and commitment to – conservation. Their very presence and continued work is often the only form of protection available to bonobos inhabiting the study sites. Foreign researchers stimulate the local economy by hiring trackers and camp attendants and by participating in local commerce. In addition, they frequently present conservation education programs in villages and distribute educational materials.

Conservation Outreach

Conservation outreach and education is a critical component to any long-term strategy aimed at protecting endangered species from over-exploitation. In parts of their range, bonobos are hunted for food and infants are sold into the pet trade. The ZSMC designed a T-shirt that is used as a conservation education tool. The shirts, available in "rain forest green" or "bonobo black," have a short conservation message in English, French, and Lingala. The message translates as: "Bonobo or pygmy chimpanzee is another kind of chimpanzee. This chimpanzee lives only in the rain forest of Zaire. Take care of the forest and watch over the bonobo."

The T-shirts are sold by the ZSMC, and proceeds fund the purchase and shipment of additional shirts to Zaire. Through researchers or expatriates living in Zaire, the shirts are distributed in Kinshasa and in remote bonobo habitat areas. To date, over 400 shirts have been distributed. In Kinshasa, the American school (TASOK) sponsors a Roots and Shoots Program, an environmental education project of the Jane Goodall Institute; the bonobo shirts donated to them served as a special fund-raising opportunity for their program. The club sells the shirts within the expatriate community and uses the funds to plant trees on denuded hillsides. While T-shirts are not a unique idea, the project has met with enthusiasm because clothing is a useful and sorely needed item. The cost of one T-shirt (US$12.00) is

Figure 1. A Tracker, working at the bonobo research camp at Wamba, Zaire, wears his "bonobo uniform." Design and distribution of these T-shirts was an important conservation education project of the Zoological Society of Milwaukee County. (Photo by Evelyn Ono-Vineberg.)

approximately equivalent to the monthly salary for game wardens (K. Ruf, personal communication), thus the T-shirts have both monetary and intrinsic conservation value. In habitat areas, the shirts have been given in exchange for conservation service and used as attendant uniforms at study sites (J. Thompson, personal communication) (Figure 1). The following message accompanied a photograph of field assistants wearing their "bonobo uniform":

> The uniforms are socially crucial in Zaire. These men tell the story that when they wear their uniforms into the forest, the bonobos will see them and know that they come as friends. The people are very much afraid of the bonobos so this uniform is significant in that, when the bonobos see it, they will be kind to the wearer. (J. Thompson, personal communication.)

Conservation Education - The *Cahier* Project

Zairian school children often lack basic school supplies. Pens, paper and books are either unavailable or in very short supply. An especially useful item is the *cahier*, a small school notebook, resembling the "exam bluebooks" sold in the U.S. Traditionally, *cahier* covers are illustrated with a picture of an animal or a regional map. In urban areas, *cahiers* are sold on the street for about 35 cents (US). The SSP and the ZSMC recently funded Delfi Messinger, founder and President of the Zoological Society of Kinshasa, to develop and publish a bonobo conservation *cahier*. Messinger has designed several *cahiers* and conservation pamphlets featuring animals of Zaire, such as the okapi, leopard, and forest elephant. The bonobo *cahier* features a colorful bonobo picture on the cover and contains natural history notes about the species (Figure 2). The plan is to make the *cahier* commercially available (as opposed to free production and distribution), so that the impact on competition within the local market is minimized. The inside text (in French and Lingala) describes the natural history of

Figure 2. Cover of the bonobo *cahier*, or school notebook, designed by Delfi Messinger, Zoological Society of Kinshasa. The Bonobo SSP and the Zoological Society of Milwaukee County jointly funded the development and publication of these items, which contain a strong conservation message. (Photo courtesy Gay Reinartz.)

the bonobo, its range and physical description, and various aspects of its behavior. The text concludes by subtly discouraging the hunting and eating of bonobos. This message is reinforced with a local folk tale of how bonobos once saved a young woman from being killed by a band of enemy warriors. In gratitude for the protection of this woman's life, her descendants do not eat bonobos. The story ends by telling that, in modern days, strangers who do not know and respect the forest come to kill the bonobo and carry off its young, but we must all cooperate to protect the bonobo. The *cahiers* are produced entirely in Zaire, and the ZSMC and the SSP bought 10,000 at wholesale to distribute to schools in the interior bonobo range areas. Half the funding for the *cahier* project came from the ZSMC and half from equal contributions from the Bonobo SSP member institutions: zoos in Milwaukee, Fort Worth, Cincinnati, San Diego, and Columbus, and the Yerkes Regional Primate Research Center. This jointly-funded project demonstrates the true collaborative nature of the Bonobo SSP and is just one of many such projects planned for the future.

Development of an Action Plan

The ZSMC's most recent contribution to bonobo conservation is the production of *The Action Plan for* Pan pansicus*: A Report on Free-Ranging Populations*.* As noted above, one persistent recommendation from conservation workshops and advisory groups, primarily the IUCN's Primate Specialist Group (PSG) and the Committee for the Care and Conservation of Chimpanzees (CCCC; now defunct), was to produce a conservation action plan. A special task force under the auspices of the PSG was formed in 1991 to advance the development of the plan [Konstant, 1992]. The CCCC and the PSG approved and appointed Nancy Thompson-Handler and Richard Malenky to write the first draft (joined by G.R. for the final version). With assistance from the PSG, Primate Conservation, Inc., and Conservation International, the ZSMC provided primary funds and resources for production and distribution of the plan. It is the first species action plan (as opposed to regional action plan) to be sponsored by the IUCN. The document is designed to serve as the principal guide to direct conservation strategies for conservationists, government agencies, and funding sources. This undertaking represents a significant collaboration among bonobo field researchers around the world. Moreover, it is an important example of how zoos and supporting societies can assist and become integral to such projects.

THE ZSMC AND SSP: COLLABORATING WITH OTHER ORGANIZATIONS

Our accomplishments also include initiatives to improve the conditions and care of bonobos in Mexican zoos. The Morelia Zoo, Michoacan, Mexico, is now an SSP member. Their participation extends the SSP commitment to a country where – until recently – international wildlife trade was not regulated by CITES. Similarly, in Europe, the EEP has assisted in the confiscation of privately held

---

* Available from the Zoological Society of Milwaukee County

bonobos (van Puijenbroeck, personal communication) and has provided the means for their re-socialization into EEP breeding groups. The Bonobo SSP frequently acts as a clearinghouse for information and we are in contact with other interested organizations, such as the Species Survival Commission/IUCN (SSC), Captive Breeding Specialist Group, Primate Specialist Group, the AZA Zairian Faunal Interest Group, the Bonobo Protection Fund, and numerous field research teams. As the conservation work of the Bonobo SSP expands, we hope to strengthen the cooperation among all these groups.

CONCLUSIONS

Direct support from the ZSMC has enabled the Bonobo SSP to further its mission and link captive programs with conservation activities in Zaire. Future plans for the ZSMC and the SSP will build on these accomplishments and the infrastructure created. For the ZSMC, plans for new projects include sponsoring activities which further the *Action Plan* recommendations, such as area surveys in Zaire and educational outreach. For the SSP, the five year strategic plan calls for the production of educational materials designed for use in the U.S. and in Zaire. Innovative ideas for conservation education in the U.S., coordinated by the Columbus Zoo, will be a major focus. Very few people are aware that this fascinating ape exists. Still fewer realize the threat of its extinction. Through careful planning and cooperation, the ZSMC and Bonobo SSP will work to enhance awareness and appreciation of this species in an effort to preserve their future.

REFERENCES

Foose, T.J.; Seal, U.S.; Flesness, N.R. Captive propagation as a component of primate conservation strategies for endangered primates. Pp. 263-299 in PRIMATE CONSERVATION IN THE TROPICAL RAIN FOREST. C.W. Marsh; R.A. Mittermeier, eds. New York, Alan R. Liss, 1987.

Kano, T. THE LAST APE: PYGMY CHIMPANZEE BEHAVIOR AND ECOLOGY. Stanford, Stanford University Press, 1992.

Kano, T.; Idani, G.; Hashimoto, C. The present situation of the bonobos at Wamba, Zaire. PRIMATE RESEARCH 10(3):191-214, 1994.

Konstant, B. Bonobo Conservation Task Force update. BULLETIN OF THE CHICAGO ACADEMY OF SCIENCES 15(1):37, 1992.

Malenky, R.K.; Thompson-Handler, N.; Susman, R.L. Conservation status of *Pan paniscus*. Pp. 362-368 in UNDERSTANDING CHIMPANZEES. P.G. Heltne; L.A. Marquardt, eds. Cambridge, MassaSusman, R.L.; Badrian, N.; Badrian, A.; Thompson-Handler, N. Pygmy chimpanzee in peril. ORYX 16:179-183, 1981.

McNeely, J.A.; Miller, K.R.; Reid, W.V.; Mittermeier, R.A.; Werner, T.B. CONSERVING THE WORLD'S BIOLOGICAL DIVERSITY. Gland, Switzerland, IUCN, 1990.

Oates, J.F. IUCN/SSC PRIMATE SPECIALIST GROUP: ACTION PLAN FOR

AFRICAN PRIMATE CONSERVATION, 1986-1990. Gland, Switzerland, Washington, D.C., IUCN/WWF, 1985.

Rabb, G. SUMMARY: SSP/EEP MASTERPLAN WORKSHOP - BONOBO, December, 1989. Antwerp, Royal Zoological Society of Antwerp, 1990.

Ralls, K.; Ballou, J. Effects of inbreeding on infant mortality in captive primates. INTERNATIONAL JOURNAL OF PRIMATOLOGY 3:491-505, 1982.

Ralls, K.; Ballou, J. Extinction: Lessons from zoos. Pp. 164-184 in GENETICS AND CONSERVATION: A REFERENCE FOR MANAGING WILD ANIMAL AND PLANT POPULATIONS. C.M. Schonewald-Cox; S.M. Chambers; B. MacBryde; L. Thomas, eds. Menlo Park, California, Benjamin-Cummings, 1983.

Reinartz, G.E. THE INCLUSION OF THE BONOBO (*Pan paniscus*) INTO THE SPECIES SURVIVAL PLAN: A PETITION. Apple Valley, Minnesota, American Association of Zoological Parks and Aquariums Conservation Office, 1987.

Reinartz, G.E. BONOBO (*Pan paniscus*) SPECIES SURVIVAL PLAN: MASTER PLAN, 1994-1996. Milwaukee, Wisconsin, Zoological Society of Milwaukee County, 1994.

Susman, R.L.; Badrian, N.; Badrian, A.; Thompson-Handler, N. Pygmy chimpanzee in peril. ORYX 16:179-183, 1981.

Susman, R.L.; Kabongo, K.M. Update on the pygmy chimpanzee in Zaire. IUCN/ SSC PRIMATE SPECIALIST GROUP NEWSLETTER 4:34-36, 1984.

Tashiro, Y. Economic difficulties in Zaire and the disappearing taboo against hunting bonobos in the Wamba area. PAN AFRICAN NEWS 2(2):8-9, 1995.

Thompson-Handler, N.; Malenky, R.; Reinartz, G. ACTION PLAN FOR *Pan paniscus*: REPORT ON FREE RANGING POPULATIONS AND PROPOSAL FOR THEIR PRESERVATION. Milwaukee, Wisconsin, Zoological Society of Milwaukee County, 1995.

World Conservation Union (IUCN) -SSC Captive Breeding Specialist Group. THE IUCN POLICY STATEMENT ON CAPTIVE BREEDING. Gland, Switzerland, IUCN, 1987.

Gay Reinartz is the Conservation Coordinator for the Zoological Society of Milwaukee County and the Species Coordinator for the AZA Bonobo SSP. She is also a PhD candidate in the Department of Biological Sciences, University of Wisconsin, Milwaukee. **Gilbert K. Boese** is the President of the Zoological Society of Milwaukee County.

APPENDIX
PRIMATE CONSERVATION RESOURCES
ON THE WORLD WIDE WEB

General sources of information on conservation, with an emphasis on primates.
(Many of these sites provide links to additional resources.)

American Society of Primatologists (ASP) .... www.asp.org/
    ASP's Conservation Page .... www.asp.org/conservation
American Zoo and Aquarium Association (AZA) .... www.aza.org/
AZA's Conservation Page .... www.aza.org/aza/conservation.html
• **Brazil FIG** .... www.aza.org/aza/advisory/brazil.htm
• **Great Ape TAG** www.aza.org/aza/advisory/ape96.htm
• **Madagascar FIG** www.aza.org/aza/advisory/madfig96.htm
• **New World Monkey TAG** .... www.aza.org/aza/advisory/newworld.htm
• **Old World Monkey TAG** www.aza.org/aza/advisory/oldwor96.htm
• **Prosimian TAG** www.aza.org/aza/advisory/prosim96.htm
• **Zaire FIG** www.aza.org/aza/advisory/zfig96.htm
Primate Information Network (PIN) .... www.primate.wisc.edu/pin/
• **PIN's Conservation Page** .... www.primate.wisc.edu/pin/conserv.html
• **PIN's** Primate Species Information .... www.primate.wisc.edu/pin/geninfo.html
Conservation International .... www.conservation.org/
Wildlife Preservation Trust International ...... www.cc.columbia.edu/cu/cerc/wpti.html
World Wildlife Fund Global (WWF) .... www.panda.org/home.htm
• **WWF** - Species Under Threat .... www.panda.org/research/underthreat/
Convention on International Trade
    in Endangered Species (CITES) .... www.ns.doe.ca/biodiversity/cites.html
Rainforest Action Network (RAN) .... www.ran.org/ran/
• **RAN** - Ecotourism Information .... www.ran.org/ran/info_center/ecotourism.html
Consortium of Aquariums, Universities and Zoos .... www.selu.com/~bio/cauz/
International Species Information System
    (listing of captive primates by species/location) .... www.worldzoo.org/
Species Survival Commission
    (link to IUCN Red List of Threatened Animals) .... www.iucn.org/themes/ssc/
World Conservation Monitoring Centre
    (link to IUCN) .... www.wcmc.org.uk/species/animals/animal_redlist.html
International Primate Protection League .... www.sims.net/organizations/ippl/
International Wildlife Education and Conservation .... www.iwec.org/
Asociación Primatológica Española etologia1.psi.ub.es/
Associazione Primatologica Italiana .... www.unipv.it/~webbio/api/api.htm
Primate Society of Great Britain .... www.ana.ed.ac.uk/PSGB/
Zoo Conservation Outreach Group .... www.selu.com/~bio/zcog/
World Resources Institute Biodiversity Page .... www.igc.apc.org/wri/biodiv/

227

Individual AZA-accredited zoos on the web.

[List includes only those zoos that specifically provide a conservation message on their web page - whether primate related or not.]

Binder Park Zoo (Battle Creek, MI) . . . . www.ring.com/zoo/binder.htm
Birmingham Zoo (AL) . . . . www.bhm.tis.net/zoo/
Busch Gardens (Tampa, FL) . . . . www.bev.net/education/SeaWorld/homepage.html
Cleveland Metroparks Zoo (OH) . . . . www.clemetzoo.com/
Cincinnati Zoo and Botanical Garden (OH) . . . . www.cincyzoo.org/
Denver Zoo (CO) . . . . www.denverzoo.org/
Detroit Zoo (MI) . . . . www.detroitzoo.org
Fort Worth Zoo (TX) . . . . www.rwnet.com/FWZoo/
Happy Hollow Zoo (San Jose, CA) . . . . www.sjliving.com/happyhollow/
Henry Doorly Zoo (Omaha, NE) . . . . www.omaha.org/zoo.htm
Lincoln Park Zoo (Chicago, IL) . . . . www.lpzoo.com/
Los Angeles Zoo (CA) . . . . www.lazoo.org/
Louisville Zoological Garden (KY) . . . . www.iglou.com/louzoo/
Lowry Park Zoo, Tampa (FL) . . . . www.the-solution.com/zoo/
Metro Washington Park Zoo
    (Portland, OR) . . . . engine.caboose.com/a1topics/portland_zoo/
Micke Grove Zoo (Lodi, CA) . . . . www.imon.com/mgzoo/
Milwaukee County Zoo and Zoological
    Society of Milwaukee County (WI) . . . . omnifest.uwm.edu/zoo/zoomain.html
Minnesota Zoological Garden (MN) . . . . www.wcco.com/community/mnzoo/
National Zoological Park
    (Washington, DC) . . . . www.si.edu/organiza/museums/zoo/nzphome.htm
National Zoological Park Conservation
    Research Center (VA) . . . . www.si.edu/organiza/centers/crc/
North Carolina Zoological Park
    (Asheboro, NC) . . . . ils.unc.edu/nczoo/zoohome.html
Philadelphia Zoological Garden's (PA) . . . . www.phillyzoo.org/pz-cons.htm
Phoenix Zoo (AZ . . . . aztec.asu.edu/phxzoo/homepage.html
Pittsburgh Zoo (PA) . . . . zoo.pgh.pa.us/
Riverbanks Zoological Park (Columbia, SC) . . . . www.riverbanks.org/
Roger Williams Park Zoo (Providence, RI) . . . . users.ids.net/~rwpz/
San Antonio Zoological Gardens & Aquarium (TX) . . . . www.sazoo-aq.org/
San Diego Wild Animal Park (CA) . . . . www.sandiegozoo.org/
San Diego Zoo (CA) . . . . www.sandiegozoo.org/Zoo/zoo.html
San Francisco Zoological Gardens (CA) . . . . www.sfzoo.com/html/home.html
Santa Anna Zoo (CA) . . . . www.santaanazoo.org/
Santa Barbara Zoological Gardens (CA) . . . . www.rain.org/~sbzoo/
Sedgwick County Zoo (Wichita, KS) . . . . www.scz.org/
Topeka Zoological Park (KS) . . . . lawlib.wuacc.edu/topeka/zoo/zoo.html
Virginia Zoological Park (Norfolk, VA) . . . . www.whro.org/cl/vazoo/index.html
Wildlife Conservation Society/Bronx Zoo (NY) . . . . www.wcs.org
Woodland Park Zoological Gardens (Seattle, WA) . . . . www.zoo.org/
Zoo Atlanta (GA) . . . . www.netdepot.com/~zooatl/

Web sites devoted to individual taxonomic primate groups or specific regions.

The Bonobo Protection Fund .... www.gsu.edu/~biossr/bpf
Mahale Wildlife Conservation
    Society (chimpanzees) .... jinrui.zool.kyoto-u.ac.jp/PAN/mwcs/mwcs.html
The Budongo Forest Project
    (chimpanzees) .... units.ox.ac.uk/departments/bioanth/budongo.html
Mountain Gorilla Protection Project ... deathstar.rutgers.edu/projects/gorilla/gorilla.html
Gorilla Conservation News .... www.selu.com/~bio/gorilla/text/conservation.html
Berggorilla & Regenwald Direkthilfe Mountain
    Gorilla and Rainforest Direct Aid .... www.kilimanjaro.com/gorilla/brd/
Partners In Conservation (gorillas) kinn.com/sweetgra/sgpfre2.html
The Mountain Gorilla Veterinary
    Project .... ourworld.compuserve.com/homepages/MGVP/
The Dian Fossey Gorilla Fund .... www.gorilla.rutgers.edu
Bushmeat Project .... goldray.com/bushmeat/
Makerere University Biological
    Field Station (MUBFS) (Uganda) .... www.usu.edu/~mubfs/index.html
Pandrillus (The Drill Rehabilitation
    and Breeding Center) .... www.icmedia.com/pandrillus/drill_monkey.html
Cercopan - Forest Monkey Rehabilitation
    and Conservation .... www.uni.edu/museum/cercopan
The Vervet Monkey Foundation .... www.enviro.co.za/
Save the Colobus Monkeys (Diani Forest) ... www.arkanimals.com/wildside/monkeys.html
Old World Monkey Taxon Advisory Group .... www.halcyon.com/gledhill/owmtag/
Orangutan Foundation International
    Conservation Program .... www.ns.net/orangutan/research.html
Lion-tailed Macaque (*Macaca silenus*) .... www.halcyon.com/gledhill/
Primate Conservation in Vietnam .... coombs.anu.edu.au/~vern/icbr.html
Wild Trade in Vietnam .... coombs.anu.edu.au/~vern/wild-trade/vietnam.html
The Douc Langur Project www-rohan.sdsu.edu/faculty/lippold1/
Golden Lion Tamarin Conservation Program .... www.si.edu/glt/
Project Tamarin - Cotton-top
    Tamarin Conservation .... www.selu.com/~bio/cottontop/
Conservation International do Brasil .... www.bdt.org.br/bdt/cibrasil
Duke University Primate Center - Research,
    Education, and Conservation (prosimians) .... www.duke.edu/web/primate/
Madagascar Fauna Group .... www.selu.com/~bio/mfg/

Taxonomic Index

Author Index

Subject Index*

* Note: Personal names in Subject Index indicate these individuals are subject of discussion; for cited publications, see Author Index on page 231.

Subject Index   245

Congo Free State. See Zaire
Congo, 16
Conservation (formerly Captive) Breeding
    Specialist Group (CBSG). See IUCN/
    Conservation Breeding Specialist Group.
Conservation education programs
  and black howling monkeys, 70
  and bonobos, 221-223
  and cotton-top tamarins, 101-104, 107-108
  and drills, 170
  and gibbons, 193
  and golden lion tamarins, 117-119, 120-
    121, 126-127
  and gorillas, 203-209
  and lion-tailed macaques, 145-146
  and prosimians, 87-88, 93
  for children, 101-104, 107-108, 203-207
  for adults, 87-88, 93, 207-209
  general, 20, 29, 34-35, 43, 52-53, 55-56,
    58-59, 67-68
Conservation ethics, 63
Conservation genetics, 66
Conservation International do Brasil, 229
Conservation International, 89, 101, 227
Consortium of Aquariums, Universities and
    Zoos, 227
Convention on International Trade in Species
    (CITES), 10, 101, 120, 151, 178, 223
Cross River National Park (CRNP), 169
Cross River State Department of Parks &
    Wildlife, 168
Cross River, 166, 168-169
Cryobiology, 66
CSBG. See IUCN Conservation Breeding
    Specialist Group.
Cupari, 11

Dallas Zoo, 56
Deforestation, xi, 7-8, 23-24, 114
DeLorme, Barbara, 211
Democratic Republic of Congo. See Zaire.
Demographic analyses, lion-tailed macaque,
    132-135
Denver Zoological Gardens, 86, 228
Detroit Zoo, 46-47, 50, 57, 228
Dian Fossey Gorilla Fund (DFGF), 37, 201-
    202, 208-210, 229
Diarama (exhibit design), 51-52
Diet for zoo primates, 5, 67, 115, 125
Discovery of new primates, xii
Disease
  ebola, 13

Herpesvirus simiae, 139
  in primates, 5-7, 45, 49, 67, 181, 190, 192
  respiratory, 6, 46
  risk of transmission from reintroduced
    primates, 89, 119-120, 145-146, 190
  Salmonella, 119
  transmission, 89
  transmission from human, 6, 49
  Valley fever, 5
  yellow fever, 69
Disney Foundation Conservation Excellence
    Fund, 106, 108
Distance Learning, 59
Douc Langur Project, 229
Drill Rehabilitation & Breeding Center
    (DRBC), 160-161, 162, 168-169, 170,
    229
Duke University Primate Center, 82, 86, 91,
    229
Duluth Zoo, 134
Dye-mark for identification, 120-121

East Africa, 15, 18
Ebola, 13
Economic aspects of conservation
  Colombia, 98-108
  Madagascar, 93
  Nigeria, 169
  Zaire, 221-222
Education. See Conservation education.
Effective population size (Ne)
  gibbons, 182-189
  lion-tailed macaques, 133
El Paso Zoo, 134
Embryo transfer, 66, 170
Endangered Species Act, 101
Enrichment. See Environmental enrichment
    for zoo primates.
Entertainment, zoos as, 20, 22
Environmental enrichment for zoo primates,
    50, 164-165
Environmental resource centers, zoos as, 74
Equatorial Guinea, 13, 16, 37, 166-167
Ethiopia, 53
European Endangered Species Programme
    (EEP), 84, 153, 218-220, 224
European Union (EU), 13, 128
Euthanasia, avoidance regarding surplus
    animals, 139
Evansville Zoo, 134
Ex situ conservation programs
  bonobos, 219